College Physics

An experimental Approach

John Long, Ph.D.

innovativeink
PUBLISHING

Cover image Shutterstock.com

innovativeink PUBLISHING

www.innovativeinkpublishing.com
Send all inquiries to:
4050 Westmark Drive
Dubuque, IA 52004-1840

Print ISBN: 979-8-3197-2903-3
eBook ISBN: 979-8-3197-2906-4

Published in the United States of America

TABLE OF CONTENTS

LAB 1 | DATA PROCEDURES

GETTING STARTED

OVERVIEW

Vernier Graphical Analysis is the application that we will use to collect, graph and analyze experimental data. We will use it to collect data using wireless sensors and then to analyze the data. In this experiment you'll learn the basic capabilities of Vernier Graphical Analysis.

EQUIPMENT

- Computer

MANUAL DATA ENTRY & CURVE FITTING

BACKGROUND

We often need to identify the mathematical relationship between two variables. An important method for determining these relationships is to fit a mathematically defined curve to experimental data. Suppose we have data for two variables – an independent variable (x) and a dependent variable (y), and that we are interested in defining their mathematical relationship. There are many ways to do this, but it is common to plot them first – to get an idea of the behavior of their relationship.

The most logical step is to see if there is a linear relationship between the variables, which gives a straight line when y is plotted against x. If the slope is positive, the variables have a direct relationship and if the slope is negative, they have an inverse relationship. For example, the figure on the left has a direct linear relationship, while the figure on the right illustrates an inverse linear relationship.

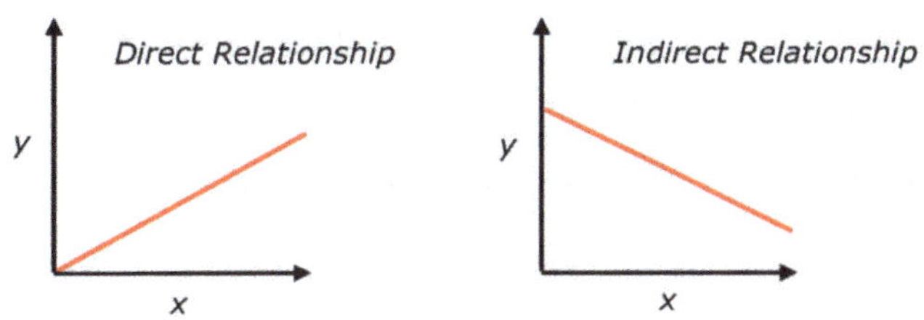

A linear equation, as you will recall from algebra, is given as: $y = mx + b$ where m is the slope and b is the y-intercept. Now, suppose we have the following graphs from our data.

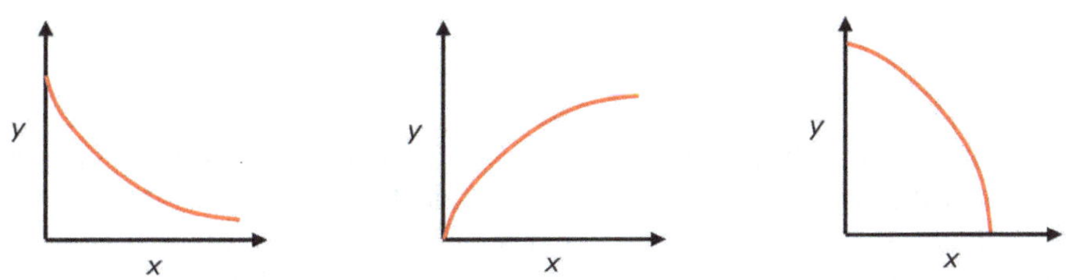

Clearly, we no longer have linear relationships. For example, the graph on the left has y decreasing as x increases, but the decrease is not proportional. Rather, y varies non-linearly with x.

Within the Graphical Analysis, we have the choice of 11 different curve families as summarized below:

Curve Family	General Equation
Proportional	$y = ax$
Linear	$y = ax + b$
Quadratic	$y = ax^2 + bx + c$
Power	$y = ax^b$
Inverse	$y = \frac{a}{x}$
Inverse Squared	$y = \frac{a}{x^2}$
Natural Exponent	$y = ae^{-cx} + b$
Natural Log	$y = a\ln(bx)$
Sine	$y = a\sin(bx + c) + d$
Cosine	$y = a\cos(bx + c) + d$
Cosine Squared	$y = a\cos^2(bx + c) + d$

a, b, c, d are constants (they can be equal to 0)

PROCEDURE

(1) Watch the Vernier Graphical Analysis Tutorial (located in D2L and on the lab computer's Desktop) for a quick overview.

(2) Read the Graphical Analyss User's Manual (also in both locations). It looks long, but it is primarily images. You're going for familiarity of capabilities, not total comprehension.

(3) Start Graphical Analysis and select Manual Entry.

(4) Enter the following data into the x and y columns for Data Set 1.

X	Y
0	2
1	3
2	4
3	5
4	6
5	7
6	8
7	9
8	10

(5) An examination of the plot tells us that this looks like a linear relationship. To get more details and to see the line, click the Graph Options button, then select Apply Curve Fit from the menu. The default curve is linear, which in this case is what we want. Note that the line goes through all the points. We could use the drop-down menu to select other relationships (e.g., proportional, quadratic, power).

(6) Click Apply. This will apply a linear fit to the data and brings up a dialog box containing some key information. This will include the formula for the curve and the correlation (r). In this example, the formula is $y = mx + b$ where $m = 1$ and $b = 2$, so $y = 1x + 2$ or $y = x + 2$.

(7) If we are using a linear relationship, we will also see the correlation r (this is the same correlation you learned in Statistics class). In this case $r = 1$. Correlation is a number between -1 and 1 and describes how closely points fit a line. The closer to ± 1, the better the fit. A positive correlation indicates the line has a positive slope and a negative correlation indicates a negative slope.

(8) If the data suggests a curved fit, then select something other than linear. Apply an inverse fit to the data we entered. As you can see, this is obviously a bad fit. Instead of r, non-linear relationships use RMSE (root mean square error). Smaller RMSEs indicate better fits (0 is as good as it gets). The RMSE for our data is 6.116, which is a terrible fit.

(9) Apply a quadratic fit. We see the RMSE is 0, which is good – in fact, it is perfect. But we already said this was a line, and this does not look like a parabola (quadratic). So, let's examine the formula: $y = ax^2 + bx + c$, where a = 7.035e-18, b = 1, and c = 2. Computer programs will often report 0 as a very small number (it has to do with limitations in floating point arithmetic) and we must be smart enough to realize it. In this case, 7.035e-18 is a very, very small number. So small in fact that it is, for all practical issues, equal to 0. Our formula becomes

$$y = ax^2 + bx + c = 0x^2 + 1x + 2 = x + 2$$

which is a linear fit with the same formula as we saw earlier.

(10) Clear the data table by highlighting all cells and then press the delete key on the keyboard or click the 3 dots next to the Data Set 1 title and select Delete Data Set.

(11) Enter the following data and fit an inverse square equation, then click Apply.

X	Y
0.5	3.000
0.8	1.170
1.3	0.444
2.0	0.188
3.0	0.083

(12) The formula and coefficients are given as:

y = a/x^2 (note x^2 means x^2)
A: 0.7499
RMSE: 0.0009396

This would give us $y = \frac{a}{x^2}$ for the general form and $y = \frac{0.75}{x^2}$ for the specific equation. Note that it is OK to round the 0.7499 to 0.75.

(13) Complete Questions 1–15.

LAB 1 – ASSIGNMENT

QUESTIONS – MANUAL DATA ENTRY & CURVE FITTING

Use the Apply Curve Fit function with the following datasets to find the general and specific equations. For the non-linear curves, try several functions – the "best" fit will be the one with lowest RMSE. *Note that the specific Equation will have numbers and the general equation will not.* None of these will use the sine or cosine functions. Report your findings in Table 1.

1.

x	y
0.6	0.198
0.8	0.264
1.5	0.495
2.0	0.660
2.5	0.825

2.

x	y
0	-1
1	3
2	11
3	23
5	59

3.

x	y
1	2
2	16
3	54
4	128
5	250

4.

x	y
2.0	1.5
2.5	1.2
1.5	2.0
3.0	1.0
5.0	0.6

5.

x	y
2.0	1.00
3.0	0.500
5.0	0.160
1.5	1.78
8.0	0.062

6.

x	y
11	0.236
17	0.064
6	1.454
28	0.0143
20	0.0393

7.

x	y
0.20	0.290
0.25	0.363
0.30	0.435
0.45	0.653
0.60	0.870

8.

x	y
0.7	3.14
1.2	3.84
1.8	5.12
4.5	21.0
2.5	8.10

9.

x	y
0.4	0.192
0.6	0.648
0.2	0.024
0.8	1.54
0.9	2.19

10.

x	y
2.5	10.00
1.5	16.67
12.0	2.083
19.0	1.316
5.0	5.000

Table 1

Problem No.	Form	General Equation	Specific Equation	RMSE
Example	Inverse Square	$y = \frac{A}{x^2}$	$y = \frac{0.75}{x^2}$	$5.52 \cdot 10^{-12}$
(1)				
(2)				
(3)				
(4)				
(5)				
(6)				
(7)				
(8)				
(9)				
(10)				

Enter the following data and use it to answer Q11.

X	Y
-3	-4
-2	-7
-1	-8
0	-7
1	-4
2	1
3	8
4	17
5	28
6	41
7	56
8	73

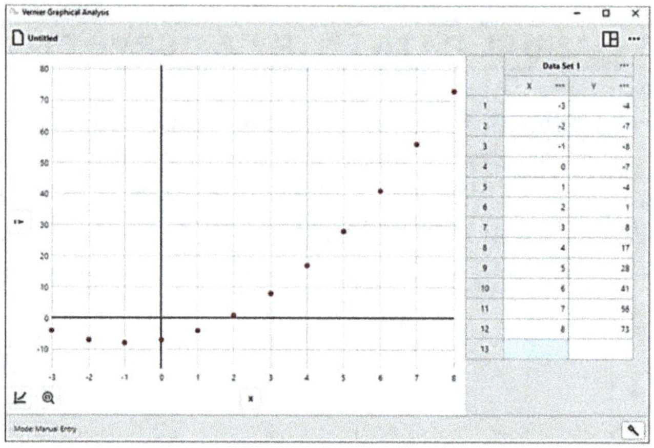

(11) What is the curve family and the <u>specific</u> formula for this data that gives the best fit? Report the RMSE as well.

(12) Under what condition (in terms of the coefficients) is an inverse squared and a power equation the same?

(13) Now let's use some "real" data. Open the file Standard Atmosphere (Click on Untitled in the upper left, then Open, then navigate to the Desktop). The file contains 860 measurements of atmospheric pressure measured in Pascals (Pa) every 100 m in altitude. We want to describe mathematically how pressure depends on altitude. Fit functions to this data until you have the "best" fitting solution. Report the type of curve, the specific formula and the RMSE.

(14) Use the specific formula from Q13 to predict the pressure at an altitude of 20 km. What is the predicted pressure and the measured pressure?

(15) What is the percent error for Q14? Percent error is: $\frac{|\text{approximate} - \text{exact}|}{\text{exact}}(100)$ and in this case, the "approximate" is the predicted value and the "exact" is the measured value.

LAB 2 | 1D KINEMATICS

GETTING STARTED

OVERVIEW

In this lab you will explore one-dimensional motion in the horizontal and vertical dimensions.

EQUIPMENT

- Lab Computer
- Dynamics Track
- Dynamics Cart
- End Stop
- Styrofoam Pad (2)
- Rubber Mat (2)
- Ruler
- Level
- Rod Stand Base
- Rod Stand (medium)
- Double Rod Clamp
- 200 g Weight with Hook
- 125 g Cart Weight (2)
- Photogate
- Photogate Accessory Rod
- Picket Fence

PT, VT & AT GRAPHS

BACKGROUND

Graphs are often the best way to describe real-world events in a compact form. There are several types of graphs for motion depending on which of the kinematic variables (time, position, velocity, acceleration) are assigned to which axis. The most common form is to plot position, velocity, or acceleration as a function of time.

PT Graphs: Position-time (PT) graphs show an object's position (y-axis) as a function of time (x-axis).

- slope is *velocity*
- the y-intercept is the *initial position*
- straight lines denote *constant velocity* while curved lines imply *acceleration*
- positive slope denotes motion in the positive direction
- negative slope denotes motion in the negative direction
- zero slope denotes a state of rest (no motion)

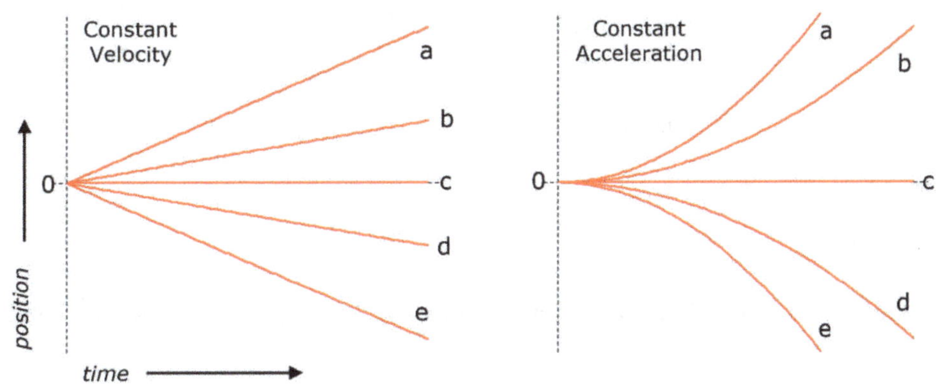

VT Graphs: Velocity-time (VT) graphs show an object's velocity (y-axis) as a function of time (x-axis).

- slope is *acceleration / deceleration*
- the y-intercept is the *initial velocity*
- straight lines denote *constant acceleration* while curved lines denote *non-constant acceleration*
- positive slope (above the x-axis) denotes speeding up
- negative slope (below the x-axis) denotes speeding up
- negative slope (above the x-axis) denotes slowing down
- positive slope (below the x-axis) denotes slowing down
- zero slope denotes motion with *constant velocity*
- the area under the curve is the *change in position*, since $d = vt$

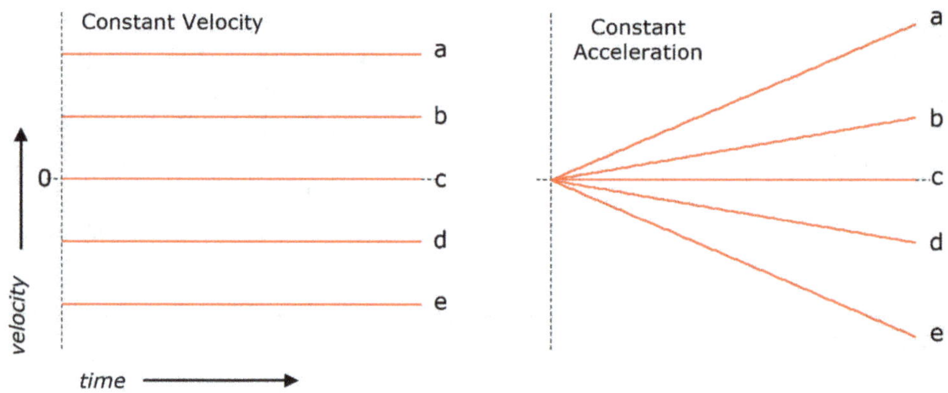

AT Graphs: Acceleration-time (AT) graphs show an object's acceleration (y-axis) as a function of time (x-axis). We won't work with these graphs; however, a summary is provided in the spirit of completeness.

- slope is *jerk*
- the y-intercept represents the *initial acceleration*
- an object undergoing *constant acceleration* traces a horizontal line
- zero slope implies motion with *constant acceleration*
- the area under the curve equals the *change in velocity*, since $v = at$

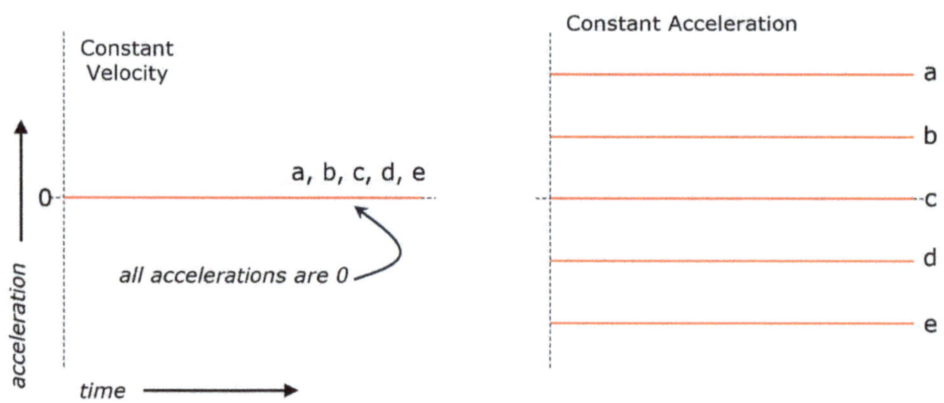

Position, time, velocity, and acceleration can be read directly off the graphs or can be calculated with data.

PROCEDURE

(1) Complete questions 1–2.

1D KINEMATIC EQUATIONS

BACKGROUND

In class, we learned the mathematical definitions of displacement, velocity, and acceleration, as well as the equations for horizontal and vertical motion. There are five kinematic variables for horizontal one-dimensional motion: time (t), displacement (Δx), initial velocity (v_0), final velocity (v_f or simply v), and acceleration (a). For vertical one-dimensional motion, the variables are: time (t), displacement (Δy), initial velocity (v_0), final velocity (v_f or simply v), and acceleration due to gravity (g).

DEFINITIONS

Time	Displacement	Velocity	Acceleration
$\Delta t = t_2 - t_1$	$\Delta x = x_f - x_0$	$v = \frac{\Delta x}{\Delta t} = \frac{x_f - x_0}{\Delta t}$	$a = \frac{\Delta v}{\Delta t} = \frac{v_f - v_0}{\Delta t}$

THE EQUATIONS OF MOTION

Horizontal	$v = v_0 + at$ (missing Δx)	$\Delta x = \left(\frac{v + v_0}{2}\right)t$ (missing a)	$\Delta x = x_0 + v_0 t + \frac{1}{2}at^2$ (missing v)	$v^2 = v_0{}^2 + 2a\Delta x$ (missing t)
Vertical	$v = v_0 + gt$ (missing Δy)	---	$\Delta y = y_0 + v_0 t + \frac{1}{2}gt^2$ (missing v)	$v^2 = v_0{}^2 + 2g\Delta y$ (missing t)

To solve a problem using the kinematic equations, you need four of the five variables. The problem will give you three of them and ask for a fourth. The strategy is to figure out the missing variable – which variable is not given, nor asked for, and using the appropriate equation as noted in the table.

PROCEDURE

(1) Complete questions 3–8.

MOTION OF A CART ON AN INCLINED TRACK

BACKGROUND

We can use various sensors to analyze motion. In this section, we'll use the sensor cart which uses an optical motion encoder system in conjunction with a sensing wheel. As the sensing wheel of the cart rotates, its position is detected by optical sensors that determine the angle turned by the wheel, which is converted to distance using the diameter of the sensing wheel.

PROCEDURE

(1) Set up the track as indicated in the photos. Set the height, h, to be approximately 15 cm from the track to the table (as measured at the front of the cart). This will allow the cart to travel to the bottom of the track in approximately 3 seconds.

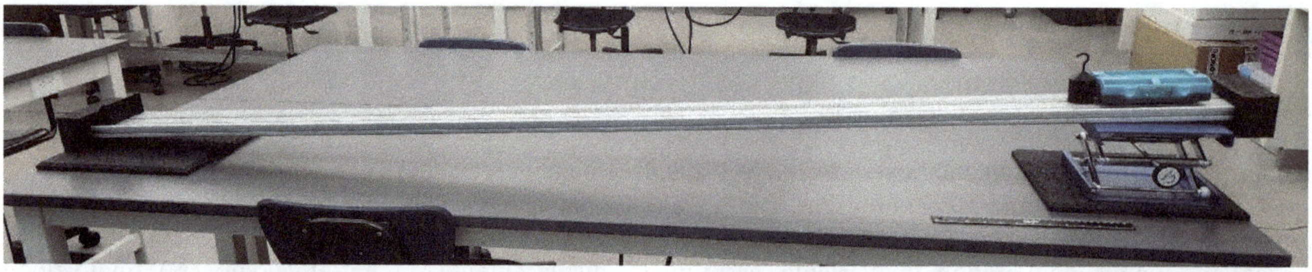

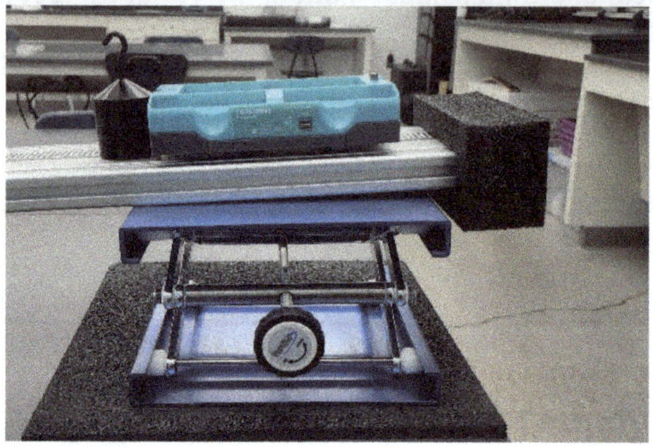

(2) Make sure the track is level (across the track) and does not tilt to the side.

(3) Enable Bluetooth on the cart by pressing the power button once (it will blink red).

(4) Launch the Vernier Graphical Analysis application.

(5) Click Sensor Data Collection.

(6) Click on the Go Direct Sensor Cart from the list of Discovered Wireless Devices. The sensor's ID is located near the barcode on the side of the cart. When connected, the Bluetooth LED will blink green.

(7) The active channel is listed in the Connected Devices Sensor Channels list. To change channels, check the box next to the Sensor Channel(s) you would like to activate. The default channel is position. Velocity and acceleration data are calculated from position information.

(8) Ensure that position is selected as the measurement channel.

(9) Click Done to enter data-collection mode.

(10) Set up the application for 3 graphs and 1 data table using the View Options Menu.

(11) The graphs should automatically be identified as Position, Velocity and Acceleration.

(12) Place the cart at the top of the ramp and place a 250 g weight in front of the cart to prevent it from moving. Ensure that the cart is oriented correctly (X + → is pointed downhill).

(13) Zero the sensor by clicking the Digital Meter in the bottom right (next to the Sensor Setup Icon), then clicking Zero. This will set the cart's current position to be the origin. The Digital Meter should read 0 m.

(14) Click the Collect Button and then release the cart (it is OK if there's a slight delay).

(15) We need to apply a curve fit to our graphs, but we need to define the region in which we want the fit to apply. In the position graph, highlight (click and drag) the area in the position plot that corresponds to smooth motion (i.e., just after the cart begins motion to just before it strikes the barrier at the end of the ramp).

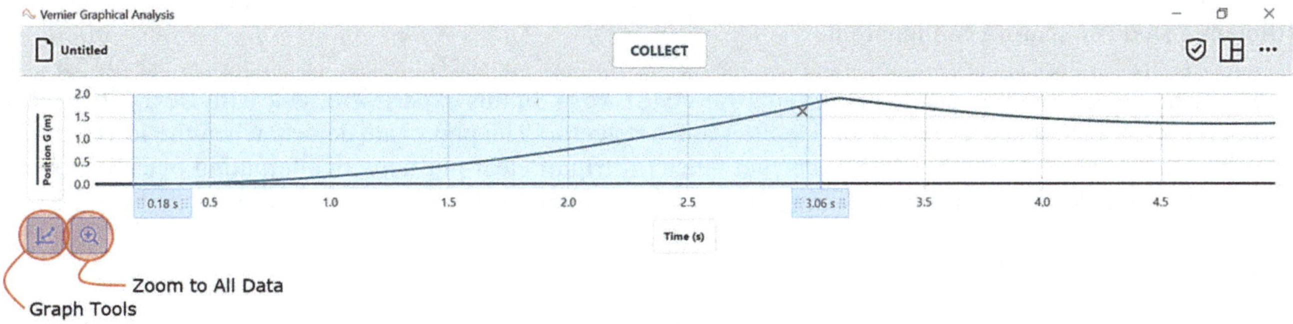

(16) Click Zoom to all Data. This should zoom to the highlighted area.

(17) Click Graph Tools – Apply Curve Fit – Quadratic. Then click Apply. A dialog box will appear, but some of it will be hidden because of space (3 graphs are too many).

(18) To fix this problem, we need to go back to View Options and select 1-graph. Now you should be able to see the whole dialog box. Record your values for a, b, and c in Table 3 (Standard Cart, Trial 1, PT Graph).

(19) Close the curve fit dialog box.

(20) Switch to the Velocity plot by clicking the label on the y-axis, selecting Velocity and unselecting Position. Click outside of the dialog box to hide it.

(21) Highlight the same region as before and fit a linear curve (straight line) to the data.

(22) Record m and b in Table 3 (Standard Cart, Trial 1, VT Graph).

(23) Repeat steps 20–22 for the AT graph. Record the data in Table 3 (Standard Cart, Trial 1, AT Graph).

(24) This completes one trial with the Standard Cart.

(25) Clear the data by clicking the Data Set Options icon (the 3 dots) and selecting Delete Data Set. You may have to scroll to the right to see the icon. Note – this is the 3 dots on the data table, NOT the big one (Other Options).

(26) Complete two more trials – DO NOT forget to zero the sensor each time before setting the cart in motion.

(27) Average your columns in Table 3.

(28) Add two hexagonal cart weights to the cart. Use the two forward (downhill) slots. The weights should fit in the slots comfortably. We will call this the Heavy Cart.

(29) Complete three trials using the heavy cart and record the data in Table 4.

(30) Turn the cart off (press and hold the Bluetooth Power button for more than three seconds). The LED indicator should stop flashing.

(31) Complete Questions 9–12.

VERTICAL MOTION – FREE FALL

BACKGROUND

An object is in *free fall* when the only force acting on it is gravity. We will assume that drag (air resistance) is sufficiently small so that we can ignore it.

We can measure acceleration due to gravity in a variety of ways. In this experiment, you will use a very precise timer and a photogate that uses an infrared emitter and a detector. The photogate detects when the infrared beam is blocked and when it is not. We'll drop a picket fence (a strip of clear plastic with alternating black bars) to block/unblock the beam. The resultant timing allows us to compute velocity and acceleration.

PROCEDURE

(1) Set up the photogate as indicated below. The entire length of the picket fence must be able to fall freely through the photogate. To avoid damaging the picket fence, make sure it has a rubber mat to land on.

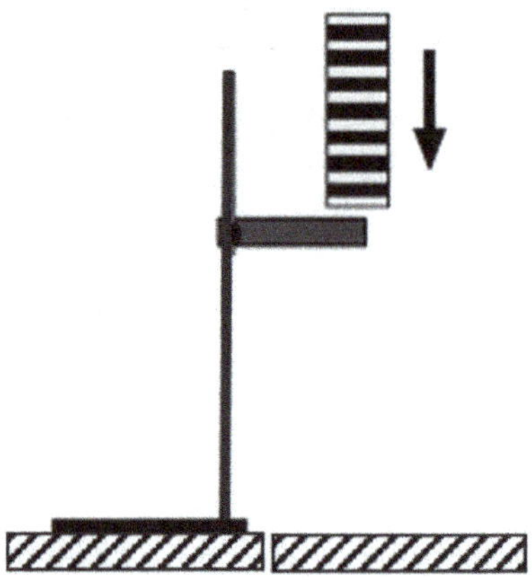

(2) Start the Vernier Graphical Analysis application.

(3) Click Sensor Data Collection and click on the photogate sensor from the list of Discovered Wireless Devices. Your sensor's ID is located near the barcode on the photogate. When connected, the Bluetooth LED will blink green.

(4) Use the dropdown menu for Sensor Channels and ensure that Velocity and Acceleration are checked. Click Done. You should have two graphs.

(5) Set up the application for two graphs and one data table using the View Options Menu.

(6) Click Collect, then drop the picket fence through the photogate. Then click Stop.

(7) Zoom to All Data on both graphs.

(8) On the velocity-time graph, apply a linear curve fit and report the slope in Table 5.

(9) On the acceleration graph, go to View Statistics, and report the statistical mean in Table 5.

(10) Repeat for a total of 5 trials.

(11) Complete Questions 13–14.

LAB 2 – ASSIGNMENT

QUESTIONS – PT, VT & AT GRAPHS

(1) Fill in Table 1 for each of the following PT graphs using the descriptors indicated for motion, direction, and velocity.

Motion = stopped, moving
Direction of Motion = stopped, right (+), or left (–)
Velocity = constant, accelerating (speeding up), decelerating (slowing down)

Table 1

	Motion	Direction of Motion	Velocity
A			
B			
C			
D			
E			
F			

(2) Fill in Table 2 for each of the following VT graphs using the descriptors indicated for direction of motion and velocity.

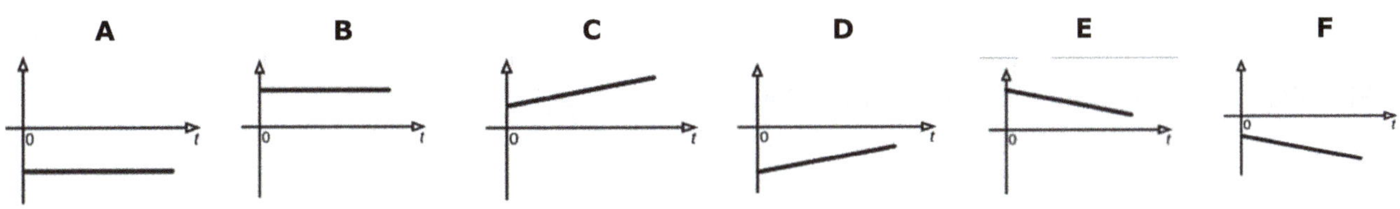

Direction of Motion = stopped, right (+), or left (–)
Velocity = stopped, constant, accelerating (speeding up), decelerating (slowing down)

Table 2

	Direction of Motion	Velocity
A		
B		
C		
D		
E		
F		

QUESTIONS – 1D KINEMATIC EQUATIONS

Plot the following data (time and position) on the graph and use it to answer Q3–7.

Time (s)	0	1	2	3	4	5	6	7	8	9	10
Position (m)	0	4	10	18	28	40	54	70	88	108	130

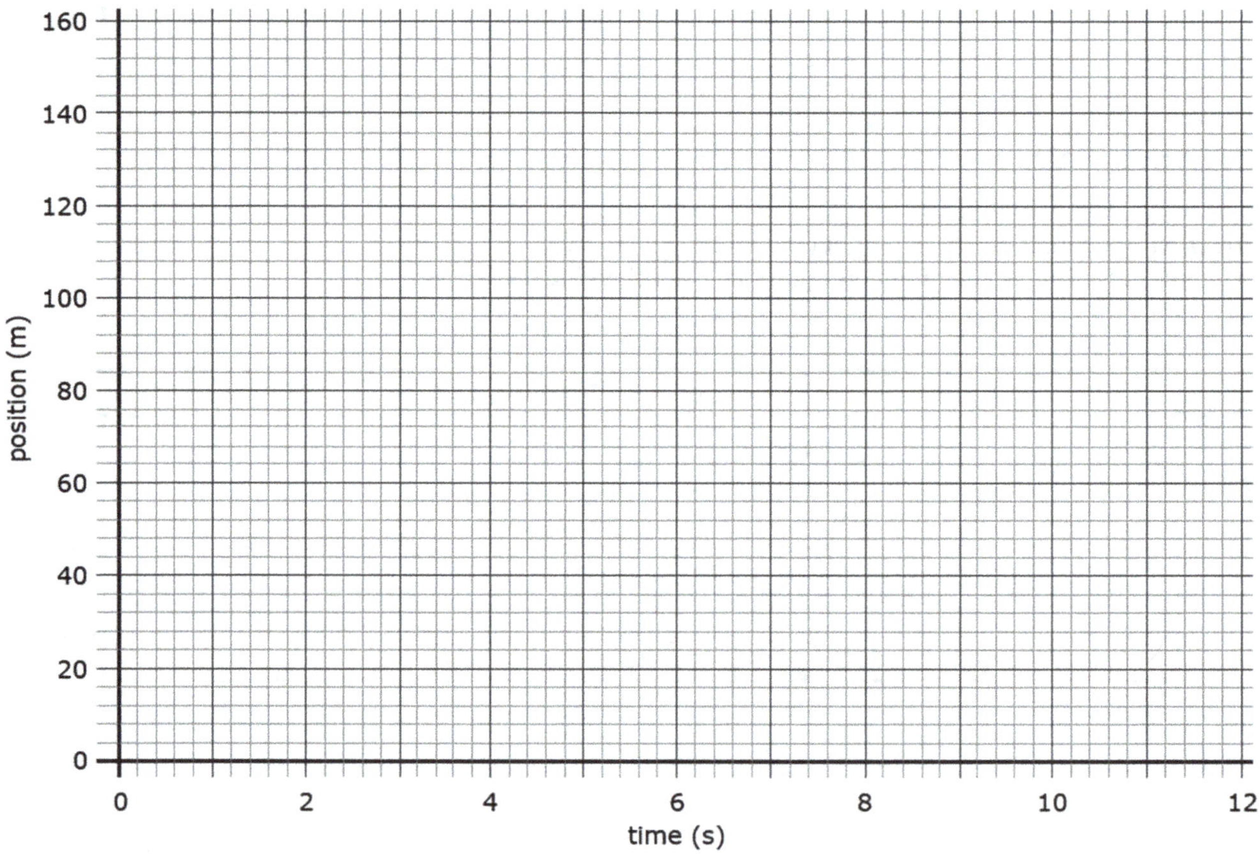

(3) Find the average velocity over the full 10 s of motion?

(4) What is the average velocity during the first 2 s of motion?

(5) What is the average velocity during the last 2 s of motion?

(6) Why are the average velocities not the same?

(7) What is the object's average acceleration over the full 10 s?

(8) A city wants to build a new airport that can handle newer and larger aircraft such as the Airbus A320. Acceleration for the A320 varies but can be as low as 2.4 m/s^2 for a fully loaded aircraft. Assuming a takeoff speed of 70 m/s, what is the shortest runway that would allow takeoff of a fully loaded A320?

QUESTIONS – MOTION OF A CART ON AN INCLINED TRACK

Enter data from your trials. Round all values to 4 decimal places in the table.

Table 3

STANDARD CART							
	Quadratic Fit (PT Graph)			Linear Fit (VT Graph)		Linear Fit (AT Graph)	
Trial	A	B	C	m	b	m	b
1							
2							
3							
Average							

Table 4

HEAVY CART							
	Quadratic Fit (PT Graph)			Linear Fit (VT Graph)		Linear Fit (AT Graph)	
Trial	A	B	C	m	b	m	b
1							
2							
3							
Average							

(9) What were the final equations (using the averaged values) for the position of each cart as a function of time?

Standard Cart:

Heavy Cart:

(10) What were the final equations (using the averaged values) for the velocity of each cart as a function of time?

Standard Cart:

Heavy Cart:

(11) Does the heavy cart move faster down the track than the standard cart? How did you determine the answer?

(12) The slope (m) of VT graph and the *y*-intercept of the AT graph both represent the cart's acceleration. Your values are likely not the same. Which one is higher?

QUESTIONS – VERTICAL MOTION – FREE FALL

Enter data from your trials. Round all values to 3 decimal places in the table.

Table 5

Trial	**1**	**2**	**3**	**4**	**5**	**Average**
Slope from VT Graph (m/s^2)						
Mean from AT Graph (m/s^2)						

(13) How well (numerically) do your values agree with one another?

(14) How well (numerically) do your values agree with local gravity of 9.805 m/s^2?

LAB 3 | VECTOR ANALYSIS

GETTING STARTED

OVERVIEW

In this lab you will work with vectors analytically, graphically, and experimentally.

EQUIPMENT

- Lab Computer
- Force table
- Center ring with attached strings
- Pulleys (4)
- Level
- Weights

VECTORS (ANALYTICALLY & GRAPHICALLY)

BACKGROUND

A vector is an object that has both a magnitude and a direction. Geometrically, we can picture a vector as a directed line segment, whose length represents the magnitude of the vector and an arrow indicating the vector's direction. We denote vectors using boldface as in **A** or **B**. When writing by hand, we cannot write boldface, so we sometimes denote vectors using arrows, such as $\vec{A}$, or sometimes these are combined as $\vec{\mathbf{A}}$.

OPERATIONS ON VECTORS

There are several operations that are valid on vectors including addition, subtraction, and scalar multiplication. A vector can be decomposed into orthogonal components and any of these operations can be applied to 1-, 2- or 3-dimensional vectors. In this course, we stick with 1D and 2D. The following diagrams are summaries of these relationships.

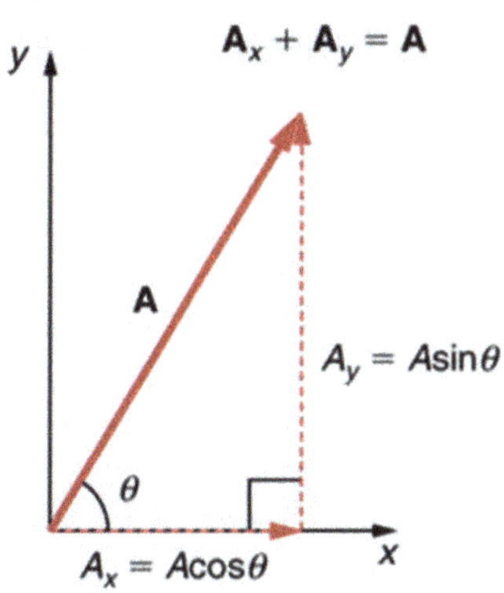

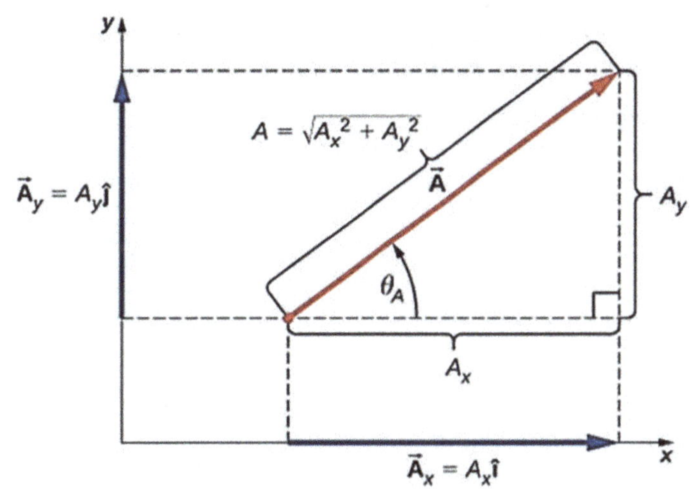

PROCEDURE

(1) Complete Questions 1–3.

VECTORS EXPERIMENTALLY (FORCE TABLE)

BACKGROUND

We can find solutions to vector problems experimentally using a force table. For any system in equilibrium, all forces that act on the system need to balance to zero. In this lab you'll experimentally find the magnitude and direction of the forces needed to place a system in equilibrium.

PROCEDURE

(1) Assemble the force table and level it by adjusting the legs if needed.

(2) Place one of the pulleys at the 0° position. Tighten the clamp well as this one will not move.

(3) Check to see that the string does not bind on the pulleys. Arrange the strings so that they radiate from a common point in the center of the ring without crossing over one another.

(4) The data tables have combinations of weights and angles with several blanks. Use the given weights and angles to find the missing ones. Determine both the angle and the weight required to keep the center ring in equilibrium (it does not touch the center pin).

(5) Complete Questions 4–5.

LAB 3 – ASSIGNMENT

QUESTIONS – VECTORS (ANALYTICALLY & GRAPHICALLY)

Use the information given to complete the tables. The units listed in the problems are Newtons (N), which is the SI unit for force (we'll learn about this soon) and represents the magnitude.

(1) Fill in Table 1 based on vector **A**. All information for a given problem is in columns.

Table 1

Problem	1	2	3	4	5
A_x	50 N	50 N	−50 N		75 N
A_y	50 N	−50 N			
A			70.71 N	90.14 N	
θ			225°	56.31°	33.69°

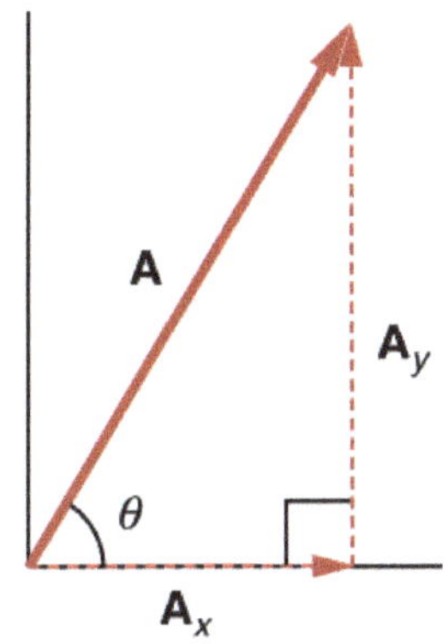

(2) Fill in Table 2 based on vectors **A**, **B**, and **R**. All information for a given problem is in columns.

Table 2

Problem	1	2	3
A_x		10 N	
A_y			
A	40 N		50 N
$θ_A$	30°		80°
B_x			
B_y		60 N	
B	10 N		30 N
$θ_B$	50°		10°
R_x		30 N	
R_y		120 N	
R		123.69 N	
θ		75.96°	

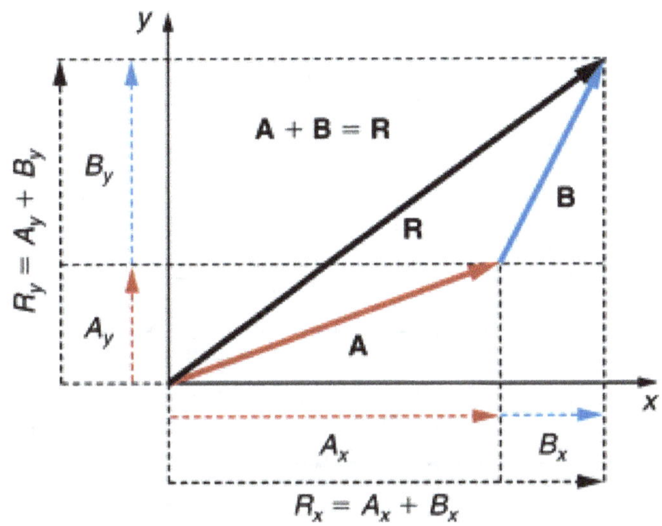

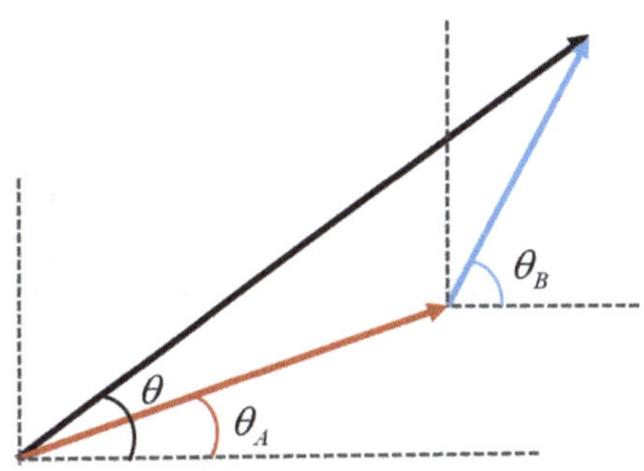

(3) An airplane flies five short segments as indicated in the diagram. Use the graphic method of vector addition and the graph below to create a scaled vector diagram to discover where the plane will end up relative to its starting position, i.e., how far and in what direction from the starting point is the plane when the segments are complete. Note that it does not matter in which order the segments are flown. All angles are relative to north (360°).

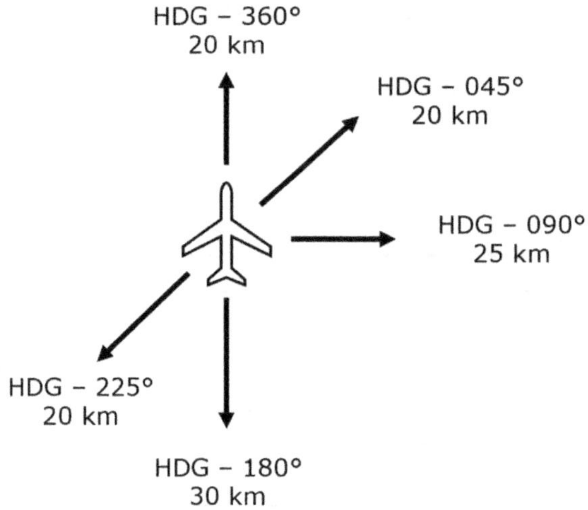

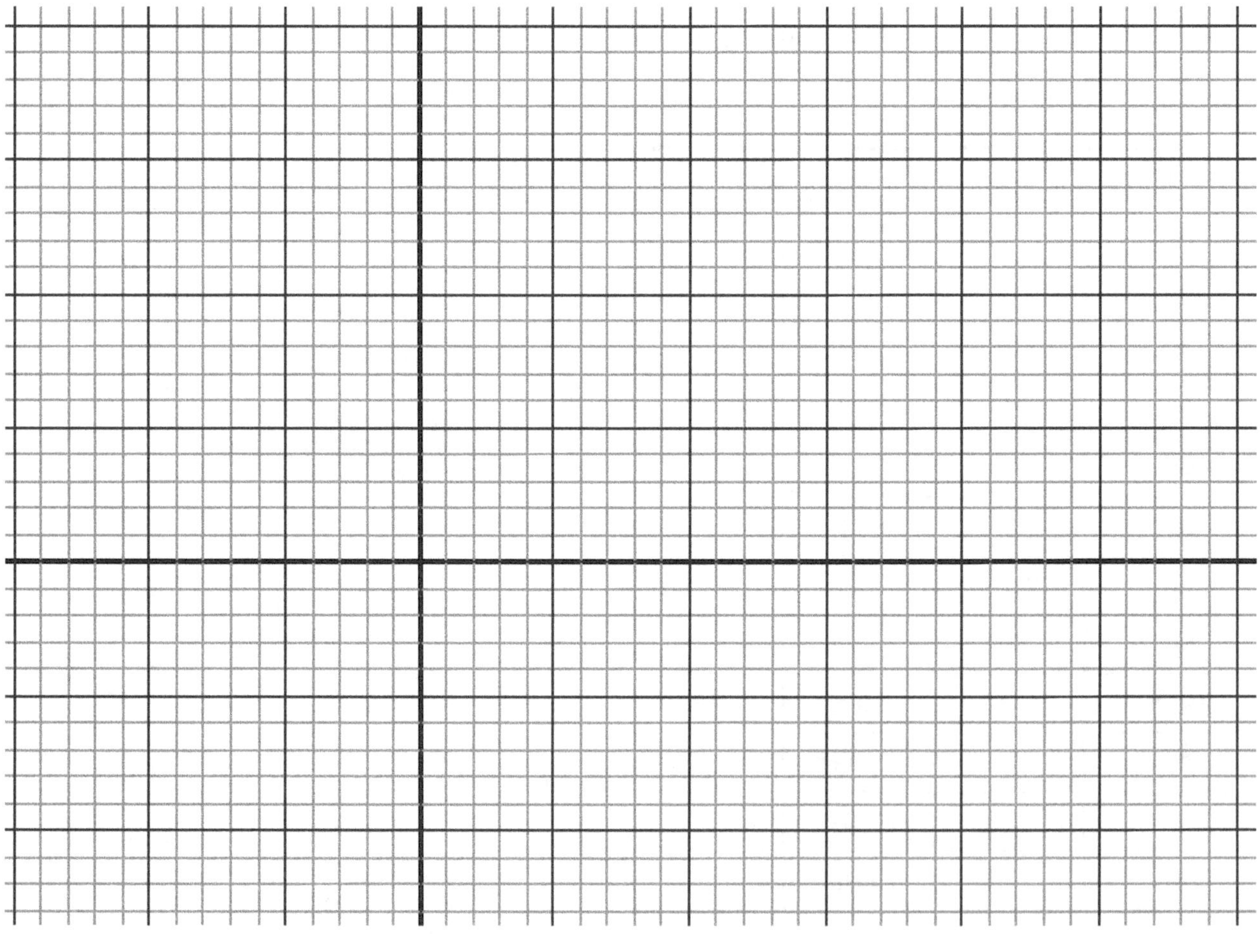

QUESTIONS - VECTORS EXPERIMENTALLY (FORCE TABLE)

(4) These problems have three or four vectors. Complete the blank cells in Table 3 with the values needed to put the system into equilibrium.

Table 3

	Vector 1		Vector 2		Vector 3		Vector 4	
Problem	**Weight**	**Angle**	**Weight**	**Angle**	**Weight**	**Angle**	**Weight**	**Angle**
1	20 g	0°	20 g	260°			----	----
2	20 g	0°	30 g	260°			----	----
3	20 g	0°	40 g	260°			----	----
4	20 g	0°	40 g	260°	40 g	140°		
5	20 g	0°	40 g	240°	40 g	160°		
6	30 g	0°	40 g	280°	40 g	210°		

(5) (a) Find the magnitude and direction of the resultant vector for $V_1 + V_2 + V_3 + V_4$? (b) What is the magnitude and direction of vector $\mathbf{V}_5$ needed for balance? Show all your work?

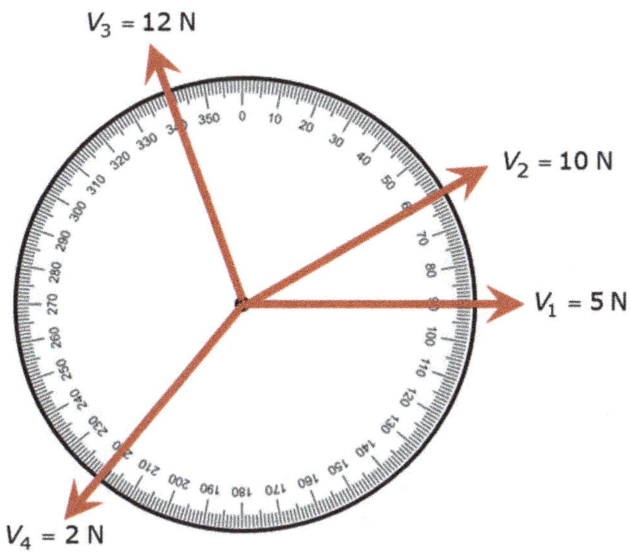

LAB 4 | 2D KINEMATICS

GETTING STARTED

OVERVIEW

In this lab you will explore motion that has a horizontal and a vertical component.

EQUIPMENT

- Lab Computer
- Rubber Mat
- Table Clamps (2)
- Projectile Launcher
- Hand Pump
- USB Power Cable
- Steel ball
- Projectile stop
- Waxed Marking Paper
- Electrical Tape
- Meter Stick
- Time-of-Flight Pad
- Launcher Cable

PROJECTILE MOTION

BACKGROUND

Projectile motion is the 2D movement of an object (the "projectile") that is launched through air. The object follows a parabolic path (due to gravity) and includes familiar objects such as a football or a bullet. At this stage of the course, we will simplify reality by ignoring air resistance.

Applying the familiar 1D equations of horizontal motion to the x- and y-directions:

1D Horizontal Motion	2D x-Direction	2D y-Direction
$v = v_0 + at$	$v_x = v_{0x} + a_x t$	$v_y = v_{0y} + a_y t$
$\Delta x = \left(\frac{v + v_0}{2}\right)t$	$\Delta x = x_0 + (v_x)_{avg}\, t$	$\Delta y = y_0 + (v_y)_{avg}\, t$
$\Delta x = x_0 + v_0 t + \frac{1}{2}at^2$	$\Delta x = x_0 + v_{0x}t + \frac{1}{2}a_x t^2$	$\Delta y = y_0 + v_{0y}t + \frac{1}{2}a_y t^2$
$v^2 = v_0^2 + 2a\Delta x$	$v_x^2 = v_{0x}^2 + 2a_x(x - x_0)$	$v_y^2 = v_{0y}^2 + 2a_y(y - y_0)$

The key to analyzing two-dimensional projectile motion is to break it into two independent motions, one along the horizontal axis and the other along the vertical. Note that, in projectile motion, acceleration in the y-direction (a_y) is gravity, g, which is a constant. Projectile motion can easily occur in 3D, but we will limit our discussion to 2D examples.

Position is given by the x- and the y-components of a parabola. The horizontal component of velocity (v_x) is constant; but, the vertical component (v_y) as well as the resultant velocity vector (v), is affected by gravity.

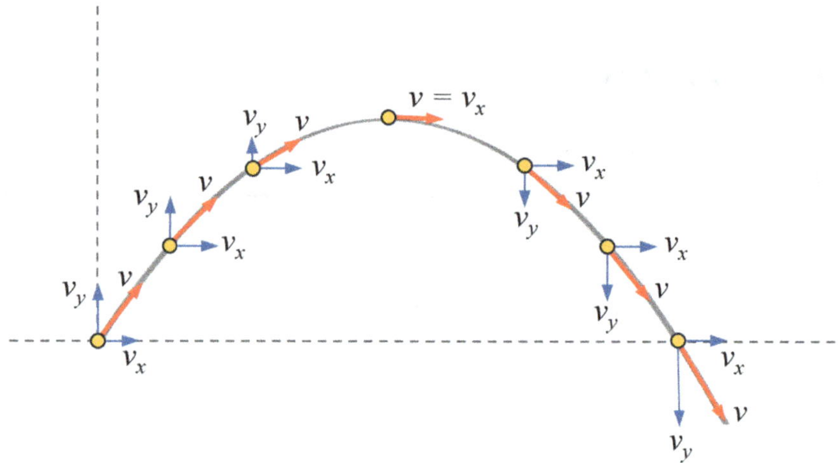

We also developed equations to find the range, maximum height, and time of flight for projectile motion

Range	$R = \dfrac{v_0^{\,2}\sin(2\theta)}{g}$
Height	$h = \dfrac{v_0^{\,2}\sin^2(\theta)}{2g}$
Time of Flight	$t = \dfrac{2v_0 \sin\theta}{g}$

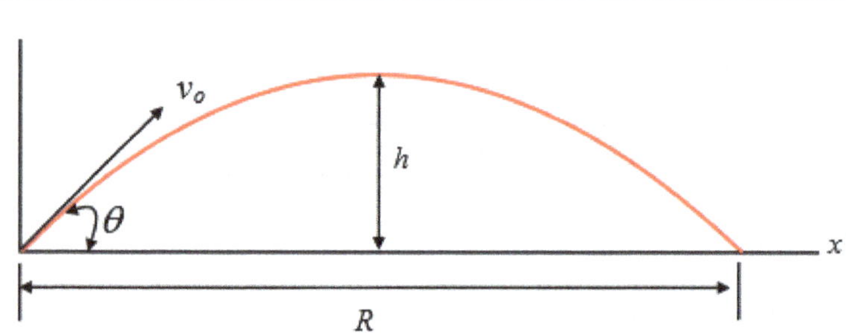

PROJECTILE LAUNCHER

The projectile launcher is used to investigate parabolic motion. It launches a steel ball at various angles and speeds. A pump generates the pressure needed to launch the steel ball. The range knob determines how far the ball will travel by controlling the maximum pressure in the launch chamber. The launch angle is reported by the software. Finally, we can find time of flight using the time-of-flight pad which measures the time between the initial launch and the ball striking the pad.

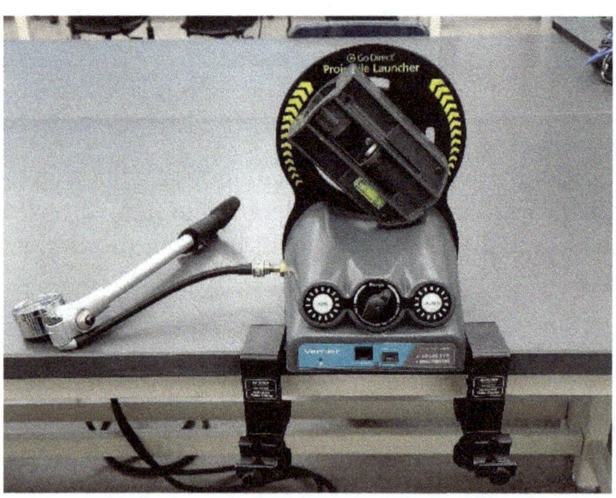

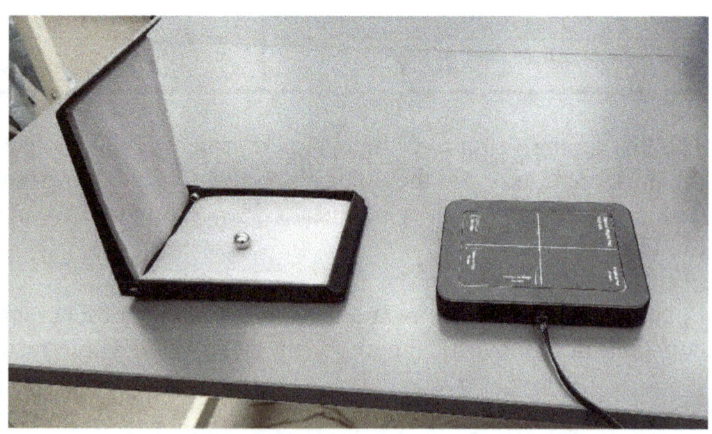

PROCEDURE

(1) Position the launcher on the edge of the table and secure it to the table with two table clamps. Ensure that the centerline of the launcher is straight down the table.

(2) The launcher requires external power. Connect the launcher via USB cable to the computer's USB port.

(3) Connect the hand pump to the launcher.

(4) Start the Vernier Graphical Analysis application.

(5) Click Sensor Data Collection.

(6) Select your launcher from the list of Discovered Wireless Devices.

> **CAUTION**
>
> - Do not put your face, hands, etc. near the end of the launch chamber while launching.
> - Do not aim the launcher at a person.
> - Do not attempt to remove the ball from the launch chamber while the chamber is pressurized.
> - Do not exceed 150 psi when pressurizing the chamber.

(7) Loosen the knob on the back of the launcher and rotate the launch chamber to the desired launch angle (45° initially), then tighten the knob. The launch angle is displayed in Graphical Analysis at the bottom right.

(8) Place a rubber mat approximately 1.5–1.9 m from the launcher (at the end of the table). This is where we want the steel ball to land.

(9) Place a projectile stop on the rubber mat and open it so that the opening faces the direction that the ball will come from.

(10) Set the pressure by adjusting the range knob on the front of the launcher. Turn clockwise (CW) for higher pressure and counterclockwise (CCW) for lower pressure. Start by turning the knob fully CW, then turn it one full turn CCW.

(11) Pump the hand pump until the pressure is above 60 psi. Then, slowly turn the range knob CCW until the pressure drops to ~40 psi. The 45° launch angle gives the maximum range, which is ~1.5–1.9 m.

(12) Ensure that the launch chamber is empty (no ball is loaded). Press and hold Arm, then press Launch. The Arm and Launch need to be pressed simultaneously for the launcher to fire.

(13) Firing the launcher uses the stored pressure. To reuse the launcher, we must reestablish the pressure. Pump the hand pump until you hear a small pressure release sound. Listen for a few small pressure release sounds and then wait five seconds to ensure the pressure has fully stabilized. You should not have to change the range knob, and the pressure should still be ~ 40 psi.

(14) Insert one steel ball into the launch chamber. The ball should be seated all the way to the back of the launch chamber.

(15) Ensure the launch path is clear.

(16) Fire the launcher (Arm + Launch). Note where the ball lands and adjust the projectile stop accordingly. If your ball lands too far, reduce the pressure slightly and try again. If your ball is too short, increase the pressure slightly and try again.

Launch Speed

(17) Many experiments involving the projectile launcher require measurement of the initial launch speed, but there is some slight variation in this value.

(18) Setup Graphical Analysis to show one data table and no graphs (recall that this is under View Options).

(19) Click Collect.

(20) Follow the basic launch procedure to launch a steel ball. The software will record various values including the launch speed. Record this in Table 1 as Trial 1.

(21) Repeat five times, so that you have a total of six trials. You must click Collect before each trial. You will get six data sets (you may have to scroll left and right to see them all).

(22) Determine the average launch speed.

Projectile Range as a Function of Launch Angle

(23) Remove the rubber mat and projectile stop from the end of the table.

(24) Place a strip of waxed marking paper (blue side down) along the centerline of the launch path from the launcher to the end of the table. Secure both ends to the table with electrical tape. The ball will leave a mark where it lands on the paper.

(25) Leave Graphical Analysis as it is (we only need it to read the launch angle).

(26) Set the launcher for 5°.

(27) Follow the basic launch procedure to launch a ball. Identify the mark left by the ball and label it with the angle. If the ball bounced on the paper and left other marks, place a small X through them so you don't get confused later.

(28) Repeat the process for the other angles listed in Table 2. Do not fill out Table 2 yet.

(29) Once you're finished, we need to measure the distance. On the tape, mark the edge of the launcher. Then measure the distance from the edge mark to each of the marks left by the ball to the nearest millimeter (which needs to be converted to meters). Since the launch point is not at the edge of the launcher, we need to add 0.082 m to each distance measurement. For example, if you measure 0.751 m, then record 0.833 m (0.751 + 0.082) in Table 2.

(30) Complete Table 2 using the following calculations for each angle. Note that v_0 is the average launch speed as determined in the first part of the experiment. Use 3 decimal places in Table 2.

- Horizontal and vertical components of velocity: $v_{0x} = v_0 \cos(\theta)$ and $v_{0y} = v_0 \sin(\theta)$
- Time of flight: $y = y_0 + v_{0y}t + \frac{1}{2}gt^2$ (where $y = 0$, $y_0 = 0.146$ m, $g = -9.8$ m/s^2), then solve for t.
- Calculated distance: $x = v_{0x}t = v_0 \cos(\theta)t$
- Experimental distance: $d = m + 0.082$

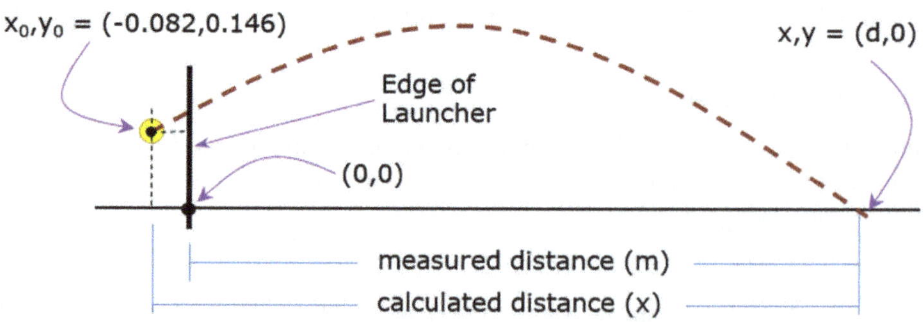

(31) Close Graphical Analysis.

Time of Flight

(32) Place the time-of-flight (TOF) pad where balls launched at a 65° launch angle landed (~ 1.5m).

(33) Use the launcher cable (the flat one) to connect the TOF pad to the accessory port on the front of the projectile launcher.

(34) Place the projectile stop near the end of the table. It should catch the steel ball after it bounces on the TOF pad.

(35) Start Graphical Analysis.

(36) It will likely detect the projectile launcher and bring up a graph. If not, click Sensor Data Collection and select your launcher from the list of Discovered Wireless Devices

(37) Click on Data Selection Setting (lower left side of Graphical Analysis). When the dialog box opens, select Mode as Photogate Timing, Measurement as Timer or Period, and check Time Between Gates. Click Done.

(38) Set the projectile launcher angle to 65°.

(39) Use View Options to select 1 graph and 1 data table.

(40) Click Collect, then launch a steel ball as before. It should land on the TOF pad and bounce into the projectile stop. Click stop.

(41) Record the time of flight (this is the 2nd Pulse Time in Table 3). Delete the data set.

(42) Repeat the process for a total of 6 trials.

(43) Compute the average time of flight.

(44) Close Graphical Analysis.

(45) Complete Questions 1–3.

LAB 4 – ASSIGNMENT

QUESTIONS – PROJECTILE MOTION

Table 1

Trial	1	2	3	4	5	6	Avg
Launch Speed (m/s)							

Table 2

θ	$V_{0y} = v_0 \sin(\theta)$	$V_{0x} = v_0 \cos(\theta)$	Time of Flight	Calculated Distance (x)	Experimental Distance (d)	Difference (d – x)
5°						
15°						
25°						
35°						
45°						
55°						
65°						

Show calculations for V_{0y}, V_{0x}, time of flight and calculated distance for 15° and 65° on the last page.

Table 3

Trial	1	2	3	4	5	6	Avg
Time of Flight (s)							

(1) For the 65° launch angle, what was your calculated time of flight (table 2), your experimental time of flight (avg from Table 3), and how close were they to each other.

(2) Use your data to calculate the height of the steel ball for $\theta = 45°$

(3) Show calculations for V_{0y}, V_{0x}, time of flight and calculated distance for 15° and 65° here. You must calculate the others, but you do not have to show your work.

15° Launch Angle	65° Launch Angle
V_{0y}	V_{0y}
V_{0x}	V_{0x}
Time of Flight	**Time of Flight**
Calculated Distance	**Calculated Distance**

LAB 5 | NEWTON'S LAWS

GETTING STARTED

OVERVIEW

In this lab, we will analyze the relationship between force, mass, and acceleration, and experimentally verify Newton's Laws.

EQUIPMENT

- Lab Computer
- Mass Set
- Dynamics Track
- Track Leveling Foot (2)
- Cart
- Cart Weights (4)
- Accessory Rod
- Ultra Pulley
- String
- Force and Accelerations Sensor (2)
- Rubber Band
- Electrical Tape

NEWTON'S FIRST LAW

BACKGROUND

Newton's First Law states that a body at rest remains at rest, or if in motion, remains in motion at a constant velocity unless acted on by a net external force. Inertia is a term to describe the tendency of an object to resist changes in its motion (or lack thereof). Consequently, Newton's First Law is sometimes known as the Law of Inertia.

PROCEDURE

(1) Place a sheet of paper so that half of it is on the table.

(2) Place a 1 kg mass on top of the paper.

(3) Try to slowly remove the sheet of paper without moving the mass.

(4) If you are successful in Step 3, let me know right away. If not, continue to Step 5.

(5) Hold the free end of the paper at a slight downward angle with one hand. Bring your other hand down on the paper with a quick strike (this will work better if the edge of your striking hand is damp).

(6) Complete Questions 1–2.

NEWTON'S SECOND LAW

BACKGROUND

Newton's First Law applies to objects for which all existing forces are balanced. If the forces are unbalanced, then Newton's Second Law applies. The second law states that the acceleration of an object is dependent upon two variables – the net force acting upon the object and its mass.

Mathematically:

$$\mathbf{F}_{net} = m\mathbf{a}$$

Remember that force and acceleration are vector quantities. This means that Newton's second law holds in all directions. Motion in one direction does not affect motion in another direction. You can always break up the forces and the acceleration into their respective components in the *x*, *y*, and *z* directions.

$$F_{net,x} = ma_x \qquad F_{net,y} = ma_y \qquad F_{net,z} = ma_z$$

In this experiment, we'll use a cart on a track attached by a string to a hanging mass.

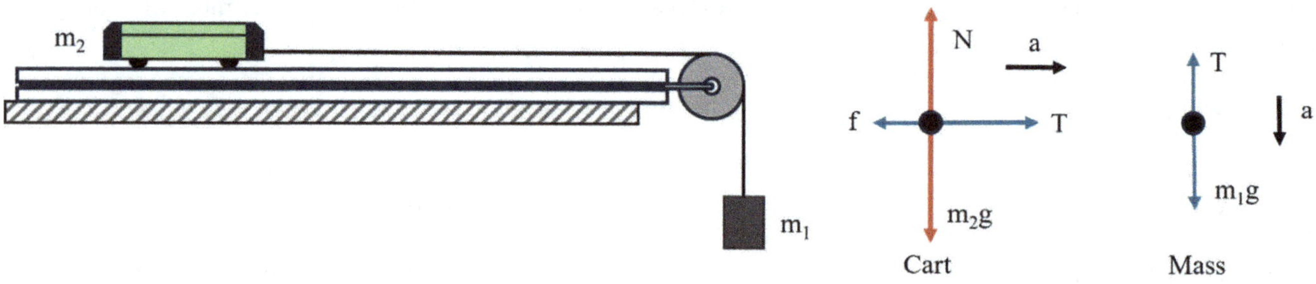

The string passes over a pulley at the end of the track and is attached to a mass. The weight of the mass provides tension in the string, which will accelerate the cart along the track. A small frictional force (f) should resist this motion, but we'll assume that it is negligible in this experiment as the pulley is nearly frictionless. We will also assume that the string is of negligible mass and there is no friction between the string and the pulley. The cart and the hanging mass will have the same magnitude of acceleration, but the direction of the acceleration will be different. The cart will accelerate to the right while hanging mass will accelerate downward.

We will take the positive direction to be in the direction of the acceleration of the two masses as shown in the free-body diagrams. Let's look at the forces acting on each mass. For the falling mass, m_1, there are no forces acting in the horizontal direction. In the vertical direction it is pulled downward by gravity giving the object weight, $W = m_1 g$ and upward by the tension T in the string. Thus, Newton's second law applied to the falling mass in the y direction will be:

$$\mathbf{F}_{net,1} = m_1 g - T = m_1 a$$

Where the downward direction has been chosen to be positive (this is counter to what we normally do in class).

The free-body diagram shows the forces acting on m_2 (the cart). There is no motion in the vertical direction; so, the net force and acceleration in the vertical direction will be zero. In the horizontal direction, the tension in the string acts in the $+x$ direction on the cart while the friction force between the cart's wheels and the surface of the track opposes this motion and acts in the $-x$ direction. Newton's second law, in the x direction is:

$$\mathbf{F}_{net,2} = T - f = m_2 a$$

Since the cart and the hanging mass are connected by the string, the accelerations of m_1 and m_2 represent the same physical quantity. The tensions are likewise the same due to Newton's Third Law. Combining equations, we can eliminate T.

$$T - f = m_2 a \quad \Rightarrow \quad T = m_2 a + f \quad \text{then}$$
$$m_1 g - T = m_1 a \quad \text{becomes}$$
$$m_1 g - (m_2 a + f) = m_1 a$$
$$m_1 g = m_1 a + (m_2 a + f)$$
$$m_1 g = (m_1 + m_2) a + f$$
$$a = \frac{m_1 g}{(m_1 + m_2)}$$

This gives us a handy equation for the acceleration of an object (m_2) due to a hanging mass (m_1).

PROCEDURE

(1) Set up a track and ensure that it is level. Position it so that the end (the one with the larger numbers) is hanging over the edge of the table a few centimeters. See the lab set up for reference.

(2) Attach a pulley to an accessory rod and insert the rod into the end of the track. The rod is a bit loose in the track, which will cause the pulley to twist in place. You can minimize this with some electrical tape.

(3) Weigh the cart empty and with 1, 2, 3, and 4 weights. Record the data in Table 1.

(4) Start the Graphical Analysis application.

(5) Click Sensor Data Collection.

(6) Select your sensor from the list of Discovered Wireless Devices.

(7) Ensure that the channel is Position (the default).

(8) In the sensor settings, reverse the direction.

(9) Configure the View for 1 graph. Change the *y*-axis to be acceleration. Click Done.

Constant Cart Mass

(10) Take a cart and extend the plunger (the silver rod at the end of the cart).

(11) Place two weights in the cart, then place the cart on the track. Block its ability to move toward the pulley with the 1 kg mass.

(12) Cut a piece of string ~ 0.6 m long and tie loops at each end.

(13) Attach one end of the string to the plunger on the cart and place the free end over the pulley.

(14) Attach a 10 g weight to the free end of the string.

(15) **READ ALL OF THIS STEP FIRST**. Click Collect to begin data collection. Remove the 1 kg mass that is blocking the cart. Stop the cart at the end of the track – before it goes crashing off the end. Click Stop.

(16) On the graph, highlight the area of motion, then View Statistics and report the mean in Table 2 as the experimental acceleration. Replicates are not necessary as the results tend to be very consistent.

(17) Replace the 10 g mass with 20 g and repeat the process. Record your average acceleration in Table 2.

(18) Repeat with a 50 g mass.

(19) Complete Table 2.

Constant Hanging Mass

(20) Replace the 50 g hanging mass with a 10 g mass. Remove the weights from the cart.

(21) Repeat steps 15 and 16. Record the acceleration in Table 3.

(22) Repeat the process with 1, 2, 3, and 4 weights in the cart. Record all data in Table 3.

(23) Complete Question 3.

NEWTON'S THIRD LAW

BACKGROUND

Newton's Third Law says that when one body exerts a force on a second body, the first body experiences a force that is equal in magnitude and opposite in direction to the force that it exerts. This is sometimes shortened to: "for every action, there is an equal and opposite reaction". Unlike the first two laws, which concern individual objects, the third law describes the interaction between objects.

PROCEDURE

(1) Turn on two force and acceleration sensors.

(2) Start the Graphical Analysis application.

(3) Click Sensor Data Collection.

(4) Select <u>both</u> of your sensors from the list of Discovered Wireless Devices. You will likely need to connect them one at a time.

(5) Click Done. The Bluetooth LEDs will blink green when connected.

(6) Use the Sensor Settings to reverse the direction for one of the sensors.

(7) Configure the view for 1 graph (Force vs. Time). The force for both sensors will show here and will be in different colors.

(8) The default channel (x-direction) for the force and acceleration sensor responds to force directed parallel to the long axis of the sensor. Place the sensors horizontally with the hooks facing each other and no force applied, then obtain a short loop of string and attach it to the hooks of the sensors. Leave a bit of slack in the line.

(9) Zero both sensors.

(10) Hold one Force Sensor in your hand and have your partner hold the other so you can pull on each other using the string as an intermediary.

(11) Click Collect to begin collecting data. Gently tug on your partner's Force Sensor with your Force Sensor. Also, have your partner tug on your sensor.

(12) Repeat as necessary – try to get a cool pattern. Sketch your graph in Figure 1.

(13) Repeat the experiment using a rubber band instead of the string. Sketch your graph in Figure 2.

(14) Complete Questions 4–5.

LAB 5 - ASSIGNMENT

QUESTIONS – NEWTON'S FIRST LAW

(1) Explain what happened in this experiment. Why were you unable to remove the paper slowly, but were able to do so with a quick strike? [HINT: Consider the concepts of velocity, acceleration, force, friction, and inertia].

(2) What advantage did the wet hand provide?

QUESTIONS – NEWTON'S SECOND LAW

Table 1

	Empty Cart	Cart + 1 weight	Cart + 2 weights	Cart + 3 weights	Cart + 4 weights
Mass (kg)					

Table 2

	Mass (kg)			Acceleration (m/s^2)		
Trial	m_1	m_2	m_T *	Experimental	Theoretical	|Diff|
1						
2						
3						

*$m_T = m_1 + m_2$

Table 3

	Mass (kg)			Acceleration (m/s^2)		
Trial	m_1	m_2	m_T *	Experimental	Theoretical	|Diff|
Cart	0.01					
Cart + 1 wt.	0.01					
Cart + 2 wt.	0.01					
Cart + 3 wt.	0.01					
Cart + 4 wt.	0.01					

*$m_T = m_1 + m_2$

(3) It is very likely that your theoretical values are higher than your experimental values and the difference column is probably similar regardless of m_1. Why?

QUESTIONS – NEWTON'S THIRD LAW

Figure 1

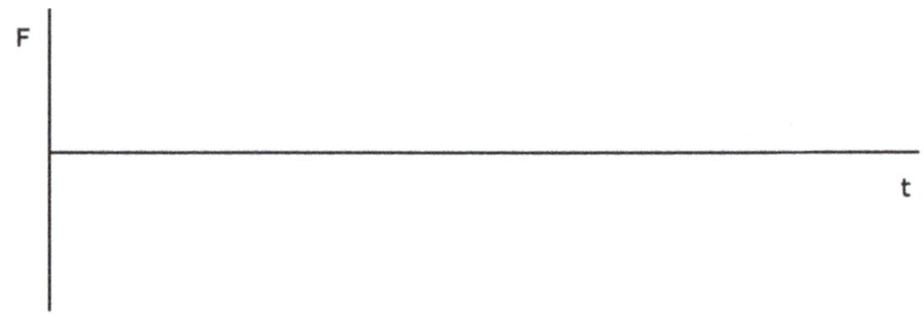

Figure 2

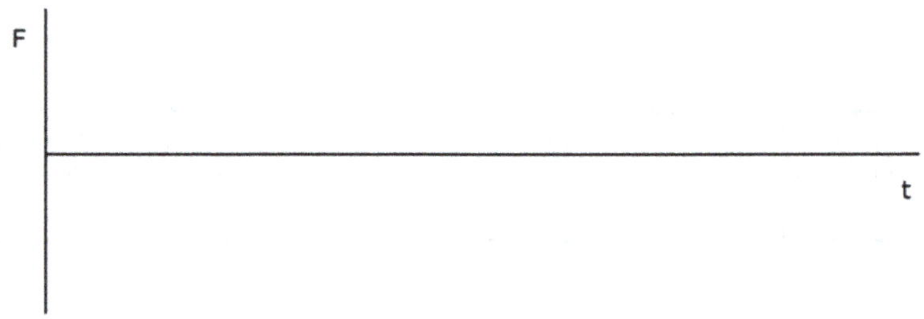

(4) Two blocks are hanging on opposite sides of a pulley. Assuming a frictionless pulley, determine the acceleration of the blocks once they are released from rest.

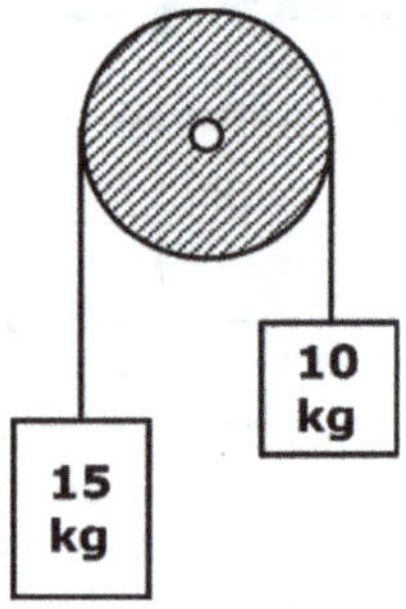

(5) A large truck and a small truck collide head-on. Use Newton's Laws to explain why it is better to be in the larger truck during the collision. Discuss in terms of velocity, acceleration, and force.

LAB 6 | RESISTIVE FORCES

GETTING STARTED

OVERVIEW

In this lab, you will experimentally determine the coefficients of static and kinetic friction and measure the effect of air resistance on a falling object.

EQUIPMENT

- Lab Computer
- Force & Acceleration Sensor
- Motion Sensor
- Friction Accessory Trays (3)
- 0.5 kg and 1.0 kg Mass
- String
- Rod Stand Base
- Rod Stand (medium)
- Lab Jack
- Wood Plank
- Angle Indicator
- Double Rod Connector (2)
- Accessory Rod (2)
- Coffee Filter

STATIC & KINETIC FRICTION

BACKGROUND

Friction is a force; therefore, it is a vector with a magnitude and direction. Friction acts in the opposite direction of an applied force. Friction is a responsive force, meaning that its magnitude varies in response to the force it is acting against.

Static friction, f_s is given by the inequality $f_s \le \mu_s N$ where μ_s is the coefficient of static friction and *N* is the normal force. Most of the time, the normal force is equal to the weight of the object. Friction acting while an object is in motion is kinetic friction and is represented by $f_k = \mu_k N$, where μ_k is the coefficient of kinetic friction.

Methodologies

We can measure friction using a force sensor, a motion sensor, or by measuring the angle at which gravity pulls an object down a slope. When using a force sensor, we measure f_s or f_k directly and calculate the normal force via mass. Then we can simply rearrange the fundamental equation to solve for friction. This method works for static and kinetic friction:

$$f_k = \mu_k N \Rightarrow \mu_k = \frac{f_k}{N} \quad \text{and} \quad f_s = \mu_s N \Rightarrow \mu_s = \frac{f_s}{N}$$

Alternatively, we can use a motion sensor to measure acceleration and use Newton's 2nd Law to measure kinetic friction.

$$F = ma \Rightarrow f_k = ma$$

Finally, we can measure static friction by finding the angle at which gravity begins to pull an object down an inclined plane. Consider a mass *m* lying on an inclined plane. If the direction of motion of the mass is down the plane, then the frictional force (*f*) will act up the plane (opposite to the motion).

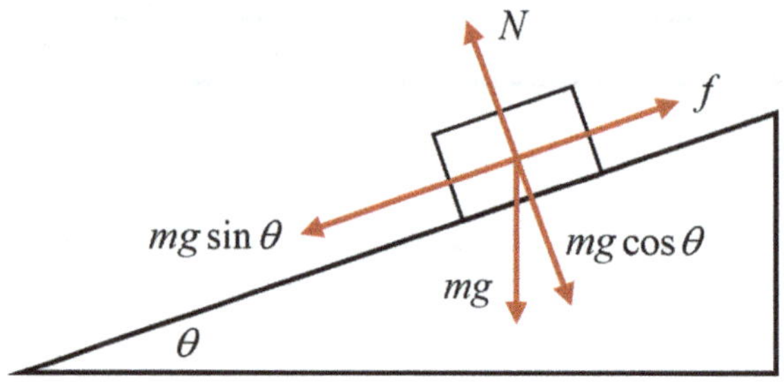

The weight of the mass is *mg* and can be decomposed into parallel and perpendicular components. Then,

- Forces up the plane = Forces down the plane: $f = mg \sin \theta$
- Forces up = forces down: $N = mg \cos \theta$

If we divide the 2 equations above, we get:

$$\frac{f}{N} = \frac{mg\sin(\theta)}{mg\cos(\theta)} = \frac{\sin(\theta)}{\cos(\theta)} = \tan(\theta)$$

and since $\frac{f}{N} = \mu$ then $\mu = \tan(\theta)$.

Therefore, we can find the coefficient of static friction by simply taking the tangent of the angle at which the object begins to slide down the plane.

PROCEDURE

Friction Measured with a Force Sensor

(1) Weigh each of the trays and record their mass in Table 1.

(2) Turn on the force & acceleration sensor and start Graphical Analysis.

(3) Click Sensor Data Collection and select your sensor from the list of Discovered Wireless Devices.

(4) Ensure that the channel is set to Force (this is the default for this sensor, so it will likely be done already). Click Done to enter Data Collection Mode.

(5) Use View Options to configure the application to display one graph and one data table.

(6) Click on Data Collection Settings. Change the End Collection setting to 10 seconds, then click Done.

(7) Use a small length of string (~10–15 cm finished length) to connect one of the trays to the hook on the force sensor. This is best done by tying small loops at each end of the string and then looping one end of the string through the other to connect to the tray, while the force sensor hook can connect directly to the other end of the string.

(8) Place the 500 g mass in the tray at the opposite end from the direction of pull (this will prevent the mass from shifting in the tray).

(9) Place the force sensor at the end of the string, but with a slight sag, so no force is exerted.

(10) Zero the sensor.

(11) Click Collect. Pull to get the tray started in motion, then continue pulling so that the tray moves with a constant velocity. You may have to do this several times to get the technique down.

(12) Mark the initial peak and record the force in Table 2 as f_k. Highlight the region of constant speed, select View Statistics and record the mean in Table 2 as f_s. See the figure below for reference.

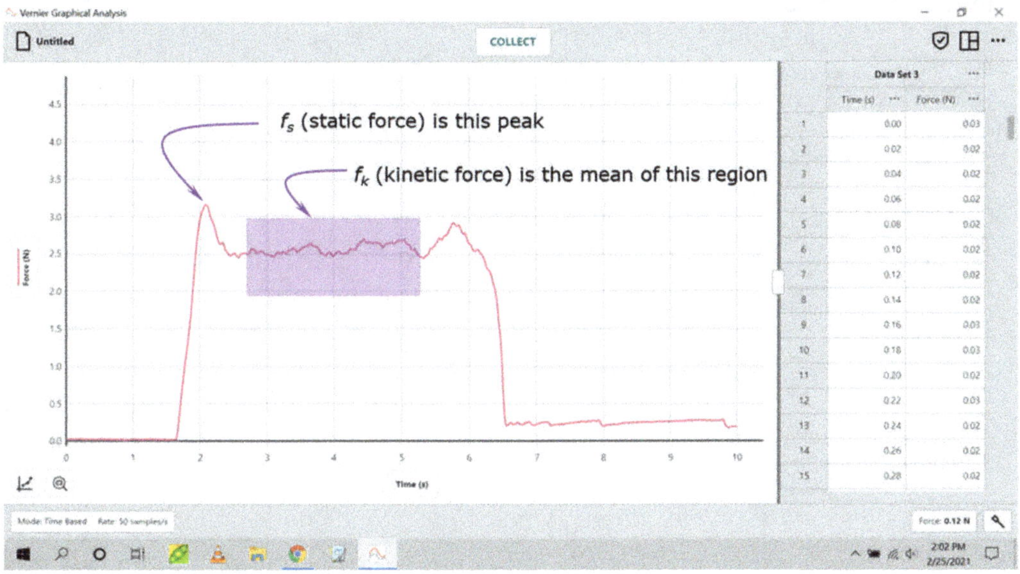

(13) Delete the data set.

(14) Repeat steps 10–13 for all combinations in Tables 2–4.

(15) Transfer the mass data from Table 1 to Tables 2–4 (make sure you're in kg). The data goes on the average line (the masses of the weights have already been placed in the tables for you).

(16) Calculate the normal force and record in Tables 2–4 ($f_N = mg$) and calculate the coefficient of static friction (μ_s) and the coefficient of kinetic friction (μ_k).

(17) Compute your grand averages.

(18) Turn off the force & acceleration sensor and close Graphical Analysis

Friction Measured with a Motion Sensor

(19) Turn on the motion sensor by pressing the power button once. The Bluetooth LED will blink red. This sensor will make an audible "click" once every second or so – this is normal.

(20) Start Graphical Analysis.

(21) Click Sensor Data Collection.

(22) Select your sensor from the list of Discovered Wireless Devices.

(23) Ensure that the channel is set to motion. Click Done to enter Data Collection Mode.

(24) Place the motion sensor at one end of the table, pointing toward the other end. Place one of the trays (empty – we will not use the weights for this part of the lab) ~ 1.5–2.0 m away.

(25) Use View Options to configure the application to display one graph and one data table.

(26) Click on Sensor Actions and reverse the direction. Zero the sensor (this marks the position of the tray as zero).

(27) Click collect (the motion sensor's "clicks" will increase in rate), then give the tray a gentle push toward the motion sensor. Minimize any movements in the zone where the motion sensor is directed.

(28) On the graph, highlight the region of "smooth" motion (e.g., see below). Apply a linear curve fit and record the slope in Table 5.s

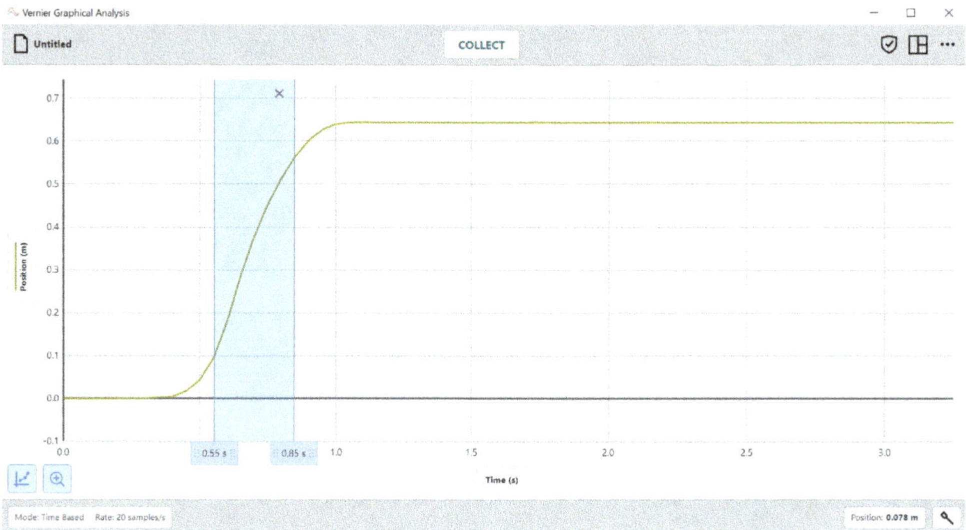

(29) Calculate the Normal force and record in Table 5 ($f_N = mg$).

(30) Using your average slope, calculate the force of kinetic friction ($f_k = ma$). Remember that slope is the acceleration. Record in Table 5.

(31) Calculate the coefficient of kinetic friction (μ_k). Record the value in Table 5.

(32) Turn off the motion sensor and close Graphical Analysis.

Friction Measured by Angle

(33) Place a lab jack at the edge of the table. Place the end of the board with the angle indicator on the lab jack.

(34) Place the tray with the plastic bottom at the top of the board (nearest the lab jack).

(35) Turn the knob to slowly raise the end of the board until the tray starts to slide.

(36) Record the angle in Table 6. Repeat twice more.

(37) Place the 500 g mass in the tray (at the front). Repeat the process for a total of three trials.

(38) Place the 1000 g mass in the tray (at the front). Repeat the process for a total of three trials.

(39) Complete Questions 1–3.

AIR RESISTANCE

BACKGROUND

Typically, when we work with falling objects, we ignore air resistance for simplicity. We assumed that there were no opposing forces in the vertical direction and that acceleration was constant. However, most objects do not fall with constant acceleration due to air resistance. Instead, they eventually reach terminal velocity. We will find the terminal velocity of a falling object (coffee filter) and use that value to calculate air resistance and the drag coefficient. There are two forces acting on a falling object: its weight, *mg*, and air resistance (drag force). At terminal velocity, the drag force is equal to the weight of the object.

Using the equation for drag force:

$$F_D = mg = \frac{1}{2}\rho C A v^2 \quad \Rightarrow \quad v = \sqrt{\frac{2mg}{\rho C A}}$$

and

$$C = \frac{2mg}{\rho A v^2}$$

m = mass of the coffee filter
g = gravity
ρ = density of air = 1.21 kg/m^3
A = area of the object = 0.15 m^2 for the coffee filter
C = drag coefficient

PROCEDURE

(1) Place a rod stand at the edge of the table. Attach a motion sensor to the stand using the double rod system. Ensure that the motion sensor is pointed downward.

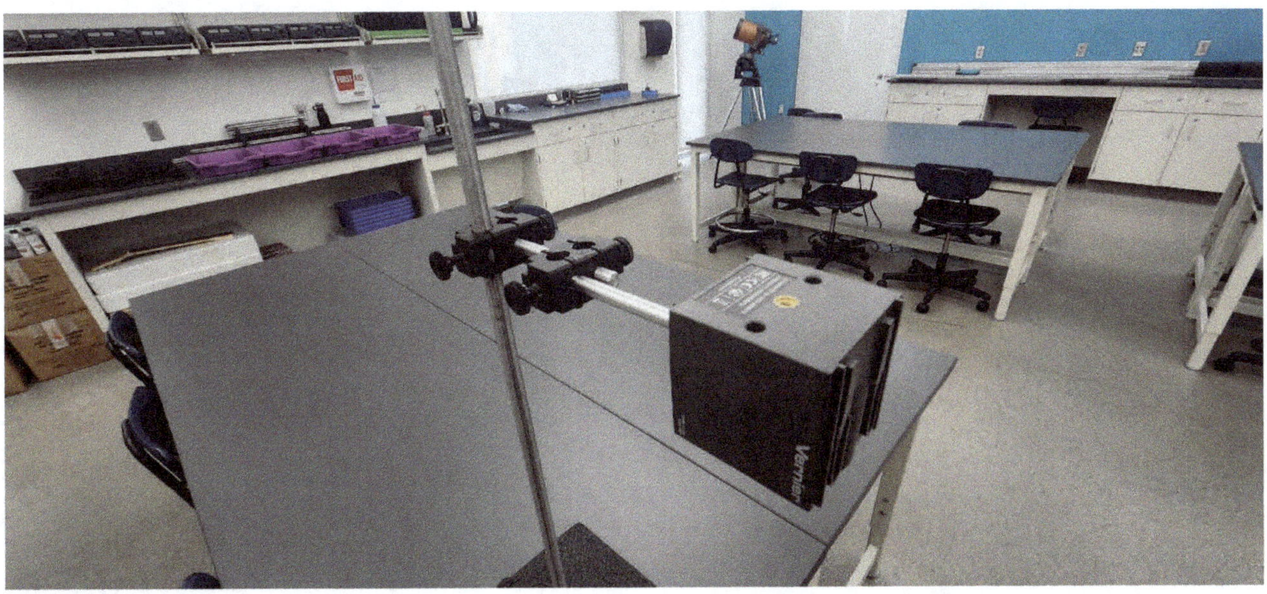

(2) Turn on the motion sensor.

(3) Start the Vernier Graphical Analysis application.

(4) Click Sensor Data Collection. Select your sensor from the list of Discovered Wireless Devices.

(5) Configure the view to show one graph (position vs. time).

(6) Weigh the coffee filter and record the mass in Table 7. Use the scale with 0.01 g precision.

(7) Place the coffee filter in the palm of your hand and hold it about 0.5 m under the motion sensor. Do not hold the filter closer than 0.15 m.

(8) Click Collect to begin data collection. When the motion detector begins to click, release the coffee filter directly below the Motion Detector so that it falls toward the floor. Move your hand out of the beam of the Motion Detector as quickly as possible so that only the motion of the filter is recorded.

(9) If the motion of the filter was too erratic to get a smooth graph, repeat the process. With practice, the filter will fall almost straight down with little sideways motion.

(10) The velocity of the coffee filter can be determined from the slope of the position *vs.* time graph. At the start of the graph, there should be a region of increasing slope (increasing velocity), and then it should become linear. Since the slope of this line is velocity, the linear portion indicates that the filter was falling with a constant or terminal velocity during that time. Highlight the linear portion of the graph and determine the slope with a linear curve fit.

(11) Record the absolute value of slope in Table 7.

(12) Repeat for a total of five trials.

(13) Turn off the motion sensor.

(14) Complete Questions 4–5.

LAB 6 – ASSIGNMENT

QUESTIONS – STATIC & KINETIC FRICTION

Table 1

Tray Bottom Material	Cork	Felt	Plastic
Mass (kg)			

Table 2 (Cork)

	Mass (kg)			Force (N)			Coeff.	
Trial	Tray	Weight	Total	Normal (f_N)	Static (f_s)	Kinetic (f_k)	μ_s	μ_k
1								
2								
3								
Avg		0.500						
1								
2								
3								
Avg		1.000						
Grand								

Table 3 (Felt)

	Mass (kg)			Force (N)			Coeff.	
Trial	Tray	Weight	Total	Normal (f_N)	Static (f_s)	Kinetic (f_k)	μ_s	μ_k
1								
2								
3								
Avg		0.500						
1								
2								
3								
Avg		1.000						
Grand								

Table 4 (Plastic)

	Mass (kg)			Force (N)			Coeff.	
Trial	**Tray**	**Weight**	**Total**	**Normal (f_N)**	**Static (f_s)**	**Kinetic (f_k)**	**μ_s**	**μ_k**
1								
2								
3								
Avg		0.500						
1								
2								
3								
Avg		1.000						
Grand								

Table 5

Material	Trial	Tray Mass (kg)	Slope (acceleration)	Normal (f_N)	Kinetic (f_k)	μ_k
Cork	1					
	2					
	3					
	4					
	5					
	Avg					
Felt	1					
	2					
	3					
	4					
	5					
	Avg					
Plastic	1					
	2					
	3					
	4					
	5					
	Avg					

Table 6

	Angle (degrees)		
Trial	**Empty Tray**	**Tray + 500 g mass**	**Tray + 1000 g mass**
1			
2			
3			
Avg (θ)			
$\mu_s = \tan(\theta)$			

(1) Which of these three methods do you feel gave you the best data? Explain.

(2) Which of these three methods do you feel was the most difficult to obtain consistent data. Why?

(3) How does your average value for the coefficient of kinetic friction (μ_k) in table 5 compare to the grand average (Tables 2–4) for each bottom material type?

QUESTIONS – AIR RESISTANCE

Table 7

Trial	1	2	3	4	5	Avg
Terminal Velocity (m/s)						
(Term. Velocity)^2 (m/s)^2						
Filter Mass (g)						

(4) Calculate F_D.

(5) Calculate the Coefficient of Drag for the coffee filter.

LAB 7 | WORK & KINETIC ENERGY

GETTING STARTED

OVERVIEW

In this lab, you will investigate the work and kinetic energy.

EQUIPMENT

- Lab Computer
- Rubber Mat
- Rod Stand Base
- Rod Stand (tall)
- Motion Sensor
- Force and Acceleration Sensor
- Accessory Rod (2)
- Double Rod Clamp (2)
- Spring Set
- Reflector
- 500 g mass
- 200 g mass
- Dynamics Track
- Motion Sensor Bracket
- Attachment Screw
- Reflector (3x5 Card)
- Spring Set
- Electrical Tape

WORK WHEN FORCE IS CONSTANT

BACKGROUND

Work is a measure of energy transfer. In the absence of friction, when positive work is done on an object, there will be an increase in its kinetic or potential energy. If the force is constant and parallel to the object's path, work can be calculated using:

$$W = |\mathbf{Fr}|\cos(\theta)$$

If the force is not constant, we could calculate the work using calculus:

$$W = \int \mathbf{F} d\mathbf{r}$$

or we can calculate the work using a graphical technique. If we divide the overall displacement into short segments, Δd, the force is nearly constant during each segment. The total work for the overall displacement is the sum of the work done over each individual segment:

$$W = \sum \mathbf{F}\Delta d$$

This sum can be determined graphically as the area under a plot of force vs. position. These equations for work can be easily evaluated using a force sensor and a motion detector.

In this part of the experiment, you will measure the work needed to lift an object straight upward at constant speed. The force you apply will balance the weight of the object, and so it is constant. The work can be calculated using displacement and the average force, and by finding the area under the force vs. position graph

PROCEDURE

(1) Turn on a motion sensor and place it on the table facing upward and turn on a force and acceleration sensor.

(2) Start the Graphical Analysis application.

(3) Click Sensor Data Collection.

(4) Select your sensor from the list of Discovered Wireless Devices. You will need to connect both sensors. Ensure the channel defaults are selected (force for the force and acceleration sensor, and motion for the motion sensor).

(5) Zero the force and acceleration sensor. Zero the motion sensor if you want, but it is not necessary.

(6) Configure the view for two graphs: force vs. time, and position vs. time. NOTE – we will need three graphs, but all three will not fit while still allowing using to do curve fits, etc. So, we will go back and forth.

(7) Hang a 200 g mass from the force sensor. The sensor should read ~1.96 N.

(8) Hold the force sensor and mass about 0.15 m above the motion sensor. Click Collect to begin data collection. Wait about 0.5 s after the clicking sound starts, and then slowly raise the force sensor and mass about 0.5 m straight upward. Hold the sensor and mass still until the data collection stops at 5 s.

(9) On the position vs. time graph, identify when the weight started to move upward at a constant speed. Record this starting time and position. Repeat for a position near the top (before the system stopped moving). If your positions are negative – it's OK, we're after the difference, so use the absolute value.

(10) On the force vs. time graph, highlight the area between the starting and stopping times identified in the previous step. Report the average force (View Statistics – mean) in Table 1 as the Exp. (experimental) Force.

(11) Use $F = mg$ to get the Calc. (calculated) Force.

(12) We need to reconfigure the view to display one graph with force on the y-axis and position on the x-axis. Highlight the area between the start and stop positions recorded in your table.

(13) From the Graph Tools, choose View Integral and report the area as the Exp. Work and use $W = Fr$ to get the Calc. Work.

(14) Repeat steps 8–13 for a total of 3 trials.

(15) Complete Questions 1–3.

WORK WHEN FORCE IS NON-CONSTANT

BACKGROUND

If the force is not constant, the previous formulas are inappropriate. In this experiment, you will measure the work required to stretch a spring and experimentally determine the spring constant. Unlike the force needed to lift a mass, the force done in stretching a spring is not constant; however, it can be calculated using the area under the force vs. position graph.

PROCEDURE

(1) Place a track on the table. Insert the motion sensor bracket into the upper slot in the track (at the starting end). Attach the first spring (green) to the bracket with the attachment screw. This will fix one end of the spring so that it does not move.

(2) Attach the reflector to the front of the force sensor.

(3) Attach the other end of the spring to the force sensor by slipping it over the hook. Position the force sensor on the track with a bit a slack in the spring. The reflector should be facing the motion sensor.

(4) Place the motion sensor on the track at the 100 cm mark.

(5) Zero both sensors and reverse the direction for the motion sensor.

(6) Configure the view for two graphs: Position (y-axis) vs Time, and Force (y-axis) vs. Position.

(7) Click Collect to begin data collection.

(8) Move the force sensor along the track towards the motion sensor. Try to keep a steady speed and don't get closer than 0.15 m to the motion sensor. It will get harder to pull the further out you get (since the force is variable). The collection will stop after 5 s.

(9) On the Position vs. Time graph, click at a point near the start of motion (but after it began). Read the time and position and record them in Table 2 as the start. Then, click at a point near the end and record these as the stop values. Here is a figure for the start (note that there was no movement until ~1.1 s).

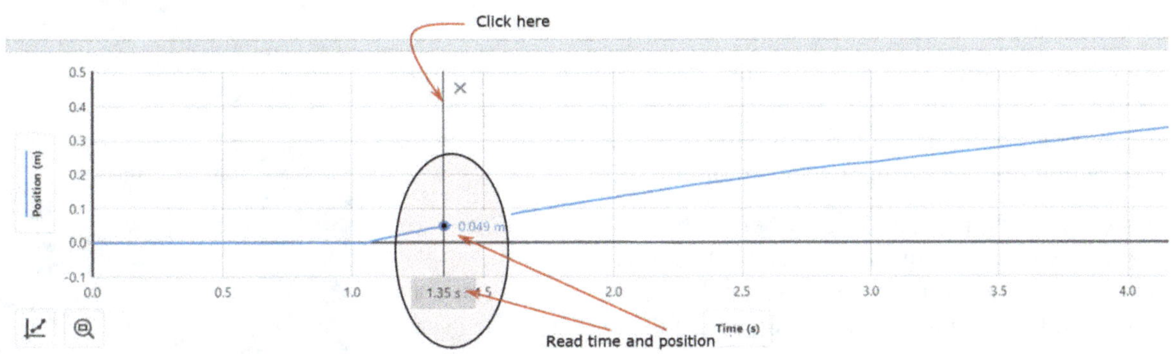

(10) On the Force vs. Position graph, highlight the region corresponding to the start and stop positions. Fit a linear curve and record the slope in Table 2 as the force constant. With the region highlighted, find the area under the curve (View Integral) and record the area in Table 2 as Work.

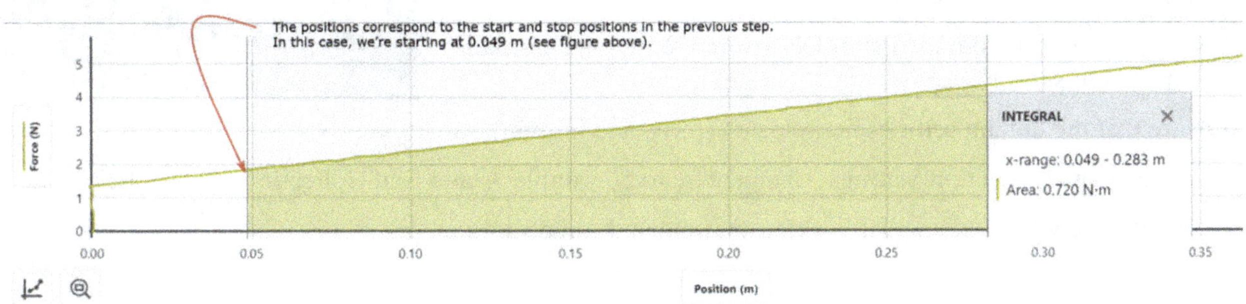

(11) Repeat for a total of 3 replicates. Be sure to zero the sensors before each trial.

(12) Swap out the green spring for the blue spring and repeat the experiment. Record your data in Table 3.

(13) Swap out the blue spring for the red spring and repeat the experiment. Record your data in Table 4.

(14) Turn off all sensors.

(15) Complete Questions 4–5.

ENERGY IN SIMPLE HARMONIC MOTION

BACKGROUND

We can describe a mass in simple harmonic motion in terms of its position, velocity, and acceleration, each as a function of time. We can also describe the system from an energy perspective. In this experiment, we'll measure the position and velocity of a mass in simple harmonic motion as functions of time, and from those data, derive the kinetic and potential energies.

There are three forms of energy in this system. Kinetic energy, $KE = \frac{1}{2}mv^2$, elastic potential energy: $PE = \frac{1}{2}ky^2$ and gravitational potential energy ($PE_g = mgy$). This form of energy can be eliminated from consideration if we measure the distances from the hanging equilibrium position. Thus, we can then concentrate on the exchange of energy between kinetic energy and elastic potential energy.

PROCEDURE

(1) Assemble the rod stand on the rubber mat. Attach a force sensor (hook downward) to the rod stand with two accessory rods (this is to place the sensor a bit further out). Then, attach the blue spring to the hook.

(2) Attach a reflector to the bottom of the 500 g mass with some tape (this helps get a good signal from the motion sensor). Attach the mass to the spring and put a bit of tape over the connection to help ensure that the mass cannot accidentally fall off.

(3) Place the motion sensor directly below the hanging mass, ensure that no extraneous objects can send reflections back to the detector. There should be approximately 0.4–0.5 m between the mass and the motion sensor. If not, adjust the height of the system as needed.

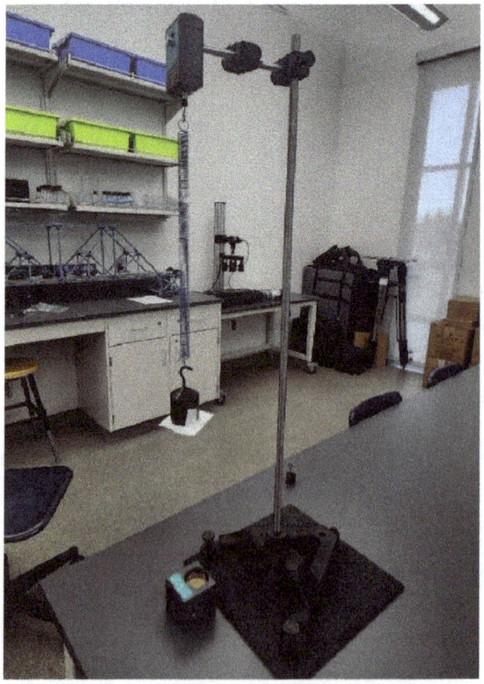

(4) Turn on the force sensor and the motion sensor.

(5) Start Graphical Analysis.

(6) Click Sensor Data Collection.

(7) Select your sensor from the list of Discovered Wireless Devices. Connect both sensors.

(8) Ensure that the default settings are selected (force and motion).

(9) Open the Data Collection Settings. Change rate to 50 samples/s and End Collection to 5 s. Click Done.

(10) Set up the view for two graphs: position vs. time and force vs. time.

(11) Make a preliminary run to verify that things are set up correctly. Lift the mass upward about 5 cm and release. The mass should oscillate along a vertical line – there should not be any horizontal motion.

(12) Click Collect to start data collection.

(13) After 5 seconds, data collection will stop and the graphs of position *vs.* time and force *vs.* time will be displayed. The PT graph should show a clean sinusoidal curve. If it has flat regions or spikes, reposition and try again. To collect again, get the mass moving, and then start data collection. Once you have a clean run, do not move the motion detector or the ring stand, as you will need this setup later.

The Relationships Between Force, Position and Velocity

(14) Click on File and choose New Experiment. Click Sensor Data Collection.

(15) Set the View for three graphs (force vs. time, position vs. time, and velocity vs. time) and one data table.

(16) Open the Data Collection Settings dialog. Change the End Collection to 10 s. Click Done.

(17) We need to measure distances from the equilibrium position of the mass, which will allow calculation of the elastic potential energy, since the distance will correspond directly to spring stretch or compression. Ensure that the mass is hanging at rest from the spring, then zero the Motion Sensor. Distances will now be measured from the current position of the mass, with displacement above the current position measured as positive, and displacement below the current position will be read as negative.

(18) Click Collect to start data collection. Start the mass oscillating in a vertical direction with an amplitude of 8–10 cm.

(19) When data collection stops, graphs of position *vs.* time and velocity *vs.* time are displayed. If it has flat regions or spikes, reposition the motion detector and try again.

(20) Carefully sketch the first 8 seconds of your graphs in Figure 1. Pay particular attention to aligning the three graphs (zoom to all data if needed). Also make sure to put values on the vertical axes (each group should be slightly different, so I did not fill them in already). The values will be easier if you click on a positive and negative peak for each graph (establishing the max positive and negative values).

Potential Energy (PE), Kinetic Energy (KE) and Total Energy (TE)

(21) Click or tap More Options (the 3 dots) in the Position column header in the table. Then, choose Add Calculated Column.

(22) Enter **PE** as the name and **J** as the units. Click Insert Expression and choose AX^B as the expression as this is the same form as $\frac{1}{2}kx^2$. Select Position as the Column for X.

(23) Multiply your spring force constant (k) from the previous experiment by 0.5 (this is $\frac{1}{2}k$). Then, enter the answer for the A value.

(24) For the B value, enter 2.

(25) Click Apply.

(26) Click or tap More Options in the Velocity column header in the table. Then, choose Add Calculated Column.

(27) Enter **KE** as the name and **J** as the units. Click or tap Insert Expression and choose AX^B as the expression as this is the same form as $\frac{1}{2}mv^2$. Select Velocity as the Column for X.

(28) Multiply your mass (500 g) by 0.5 (this is $\frac{1}{2}m$). Then, enter the answer (250 g) for the A value.

(29) For the B value, enter 2.

(30) Click Apply.

(31) Click or tap More Options in the KE column header in the table. Then, choose Add Calculated Column.

(32) Enter **TE** as the name and **J** as the units. Click or tap Insert Expression and choose AX+BY as the expression. Select KE as the Column for X and PE as the column for Y.

(33) Enter 1 for both the A and B values.

(34) Click Apply.

(35) Configure the View for three graphs: PE vs. time, KE vs. time, and TE vs. time.

(36) Carefully sketch the first 8 seconds of your graphs in Figure 2. Pay particular attention to aligning the three graphs. As before, be sure to put values on the vertical axes.

(37) Complete Questions 6–11.

LAB 7 – ASSIGNMENT

QUESTIONS – WORK WHEN FORCE IS CONSTANT

Table 1

	Time (s)		Position (m)			Force (N)		Work (J)	
Trial	**Start**	**Stop**	**Start**	**Stop**	**r ***	**Exp.**	**Calc.**	**Exp.**	**Calc.**
1									
2									
3									
Average									

*r = |start – stop|

(1) How well does your experimentally determined force compare to the calculated force?

(2) How well does your experimentally determined work compare to the calculated work?

(3) The diagram shows the result from an experiment in which an object is moved over a 10 m distance. How much work was done to move the object from 1.5 m to 8.3 m?

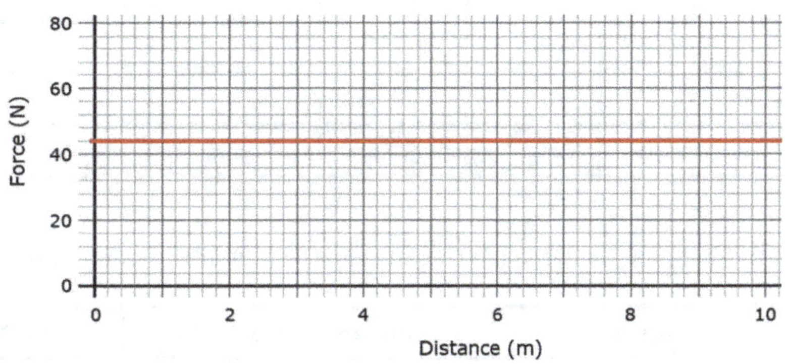

QUESTIONS - WORK WHEN FORCE IS NON-CONSTANT

Table 2

Green Spring (Force Constant = 40 ± 4 N/m)						
	Time (s)		Position (m)		Force Constant (N/m)	Work (J)
Trial	Start	Stop	Start	Stop	(Slope)	(Area)
1						
2						
3						
Average						

Table 3

Blue Spring (Force Constant = 20 ± 2 N/m)						
	Time (s)		Position (m)		Force Constant (N/m)	Work (J)
Trial	Start	Stop	Start	Stop	(Slope)	(Area)
1						
2						
3						
Average						

Table 4

Red Spring (Force Constant = 10 ± 1 N/m)						
	Time (s)		Position (m)		Force Constant (N/m)	Work (J)
Trial	Start	Stop	Start	Stop	(Slope)	(Area)
1						
2						
3						
Average						

(4) Compute the percent error for the experimental determination of the spring's force constant.

Spring	Accepted Value (N/m)	Exp. Value (N/m)	Percent Error
Green	40.00		
Blue	20.00		
Red	10.00		

(5) Consider the following force vs. position graph. The force represented is non-constant and follows the formula: $F = \frac{10}{r+1}$ where r is position in meters. Calculate the work done on the system by finding the area under the curve. There are at least 4–5 ways to do this. Pick one. SHOW ALL OF YOUR WORK.

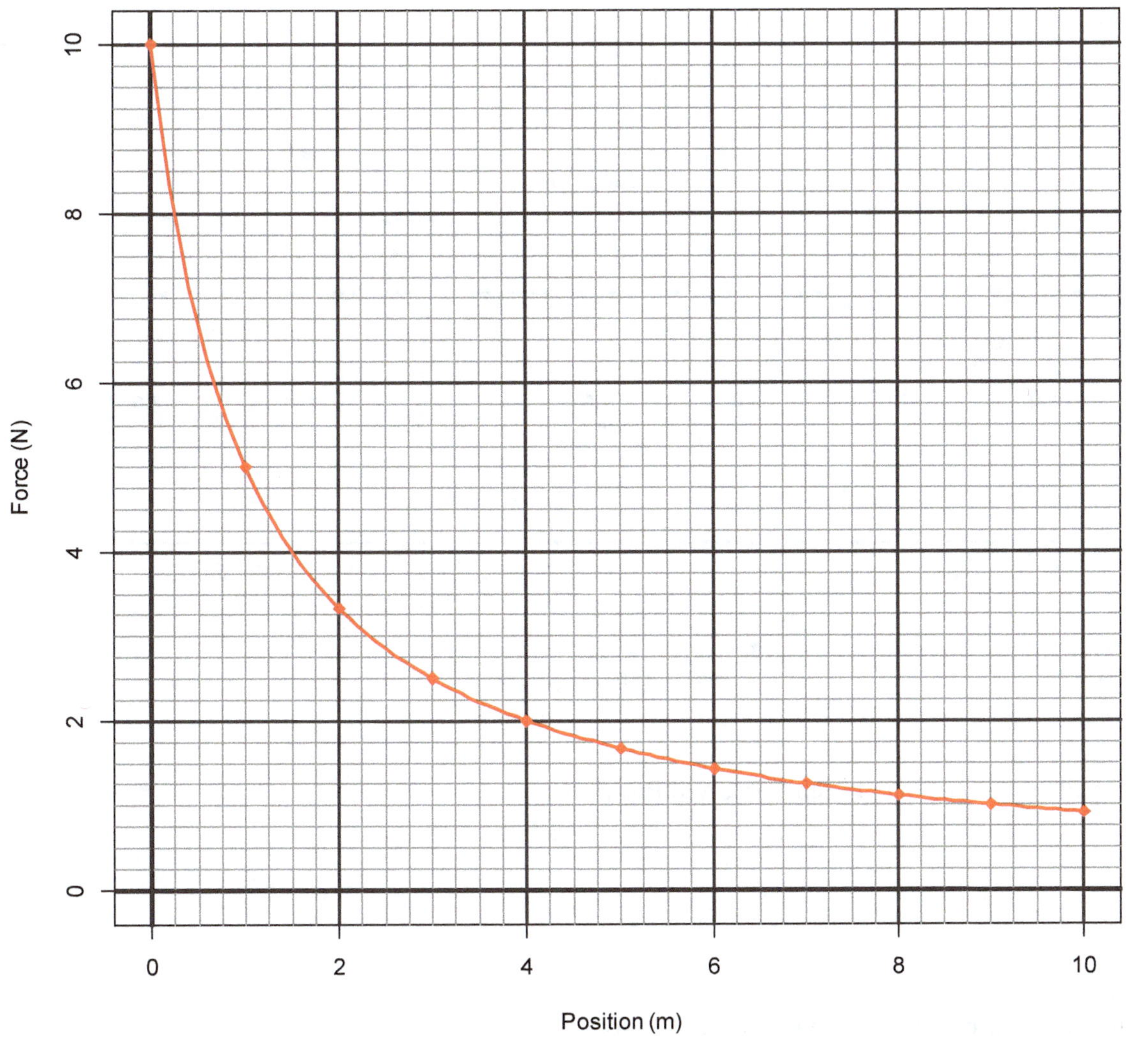

QUESTIONS – ENERGY IN SIMPLE HARMONIC MOTION

Figure 1

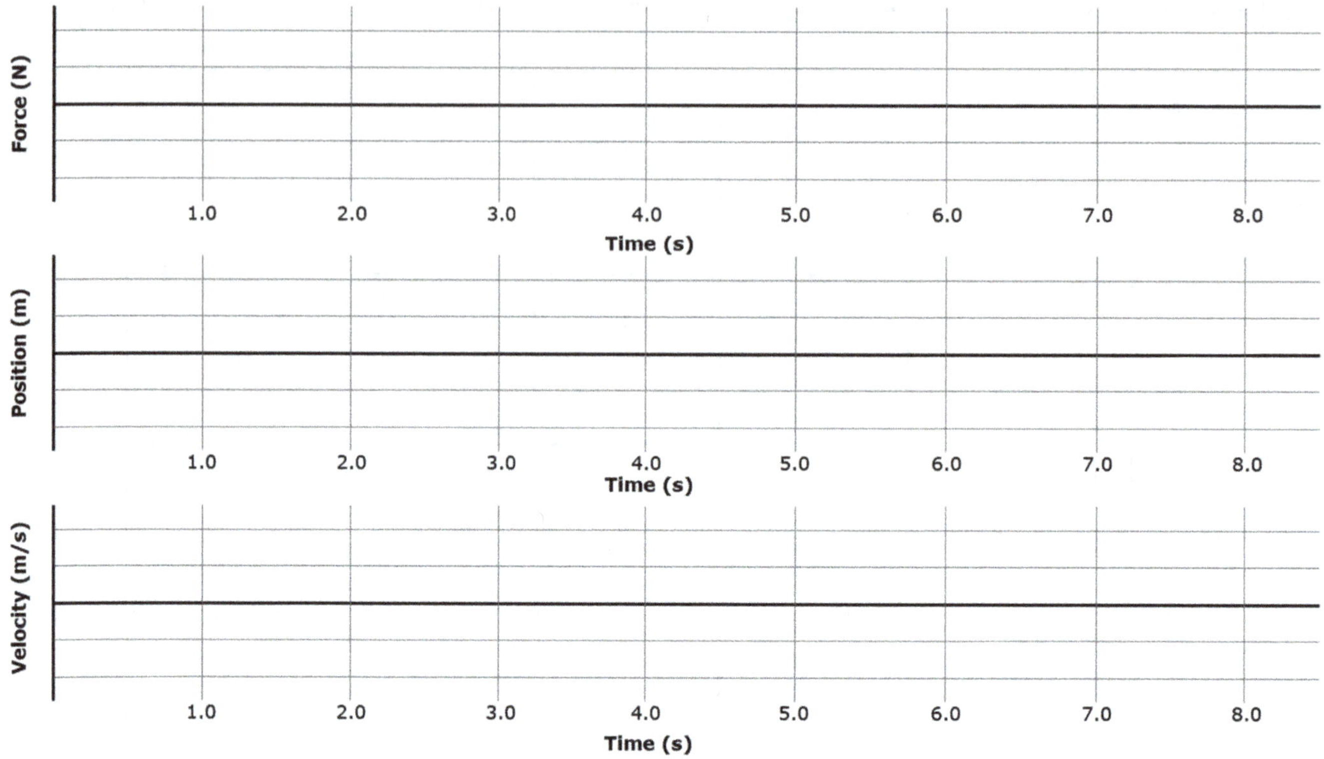

(6) Describe the relationship between Force and Position.

(7) Describe the relationship between Force and Velocity.

(8) Describe the relationship between Velocity and Position.

Figure 2

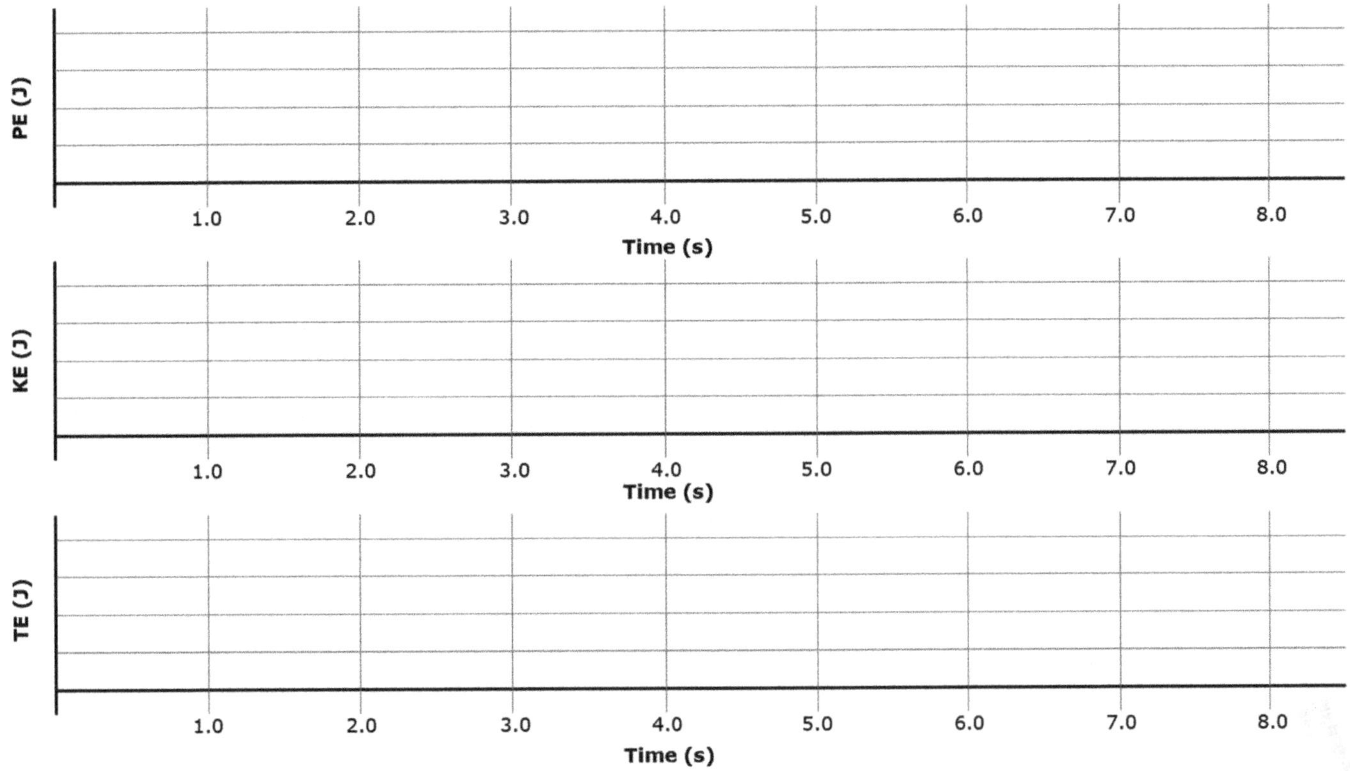

(9) Describe the relationship between PE and KE.

(10) Was energy conserved in this experiment? How do you know?

(11) A baseball player slides to a stop on level ground. The 65 kg player is running at 6 m/s at begins his slide 3.1 m from 2nd base. The force of friction, *f*, is a constant 450 N and acts over a distance of *d*. Using energy concepts, will the player make it to the base or will he slide to a stop before reaching "safety". If he stops short, how short will he be? If he makes it to base, how fast will he be moving?

LAB 8 | COLLISIONS, MOMENTUM, & IMPULSE

GETTING STARTED

OVERVIEW

In this lab, you will explore inelastic and elastic collisions in terms of Conservation of Momentum as well as the relationship between momentum and impulse.

EQUIPMENT

- Lab Computer
- Dynamics Track
- Leveling Foot (2)
- End Stop
- Styrofoam Stop (2)
- Level
- Sensor Cart (2)
- Hook and Pile Bumpers
- Magnetic Bumpers
- Hoop Bumper
- Rubber Bumper
- Play-Doh

MOMENTUM, ENERGY & COLLISIONS

BACKGROUND

In any collision, total momentum is always conserved if there is no net external force. However, this is not the case for the conservation of kinetic energy. As noted in class, collisions can be *elastic* (KE is conserved), *inelastic* (KE is lost) or *completely (or perfectly) inelastic* (the maximum amount of KE is lost). In this part of the experiment, you can observe elastic and inelastic collisions and test for the conservation of momentum and of energy.

The collision of two carts on a track can be described in terms of momentum conservation and energy conservation. As long as we do not introduce a net external force, we can expect the total momentum of the system to be conserved. In contrast, kinetic energy is not always conserved since energy is lost due to conversion to other forms such as sound, heat, etc.

Momentum is the product of mass and velocity. Velocity is a vector; thus, so is momentum:

$$\mathbf{p} = m\mathbf{v}$$

As well as the momentum of a single object, we can consider the momentum of a combination of related objects, which we call a "system".

$$\mathbf{p} = \mathbf{p}_1 + \mathbf{p}_2 + \ldots + \mathbf{p}_n = m_1\mathbf{v}_1 + m_2\mathbf{v}_2 + \ldots + m_n\mathbf{v}_n$$

The quantity $\mathbf{p}$ is conserved when there is no net external force on the system. In a collision in which two separate masses stick together after the collision and act as a single mass:

$$m_1\mathbf{v}_1 + m_2\mathbf{v}_2 = (m_1 + m_2)\mathbf{v}$$

PROCEDURE

(1) Place the track on the table. Connect a leveling foot at each end, then ensure that the track is level. Adjust the feet as necessary. Don't forget to level across the track as well as along the track.

(2) Attach an end stop at the end of the track and place a Styrofoam bumper at each end.

(3) You should have a green and a yellow cart. They are the same except for the color, which we use to help us easily identify carts. We will designate the green one as Cart 1 and the yellow as Cart 2.

(4) Set up the carts by inserting the hook-and-pile bumpers so they will be facing each other when both **+x** arrows are pointing in the same direction.

Hook & Pile (Velcro) Bumpers

Magnetic Bumpers

(5) Measure the masses of the carts and record the values in Table 1.

(6) Start Graphical Analysis.

(7) Click Sensor Data Collection. Select the sensors from the list of Discovered Wireless Devices. You will need to connect both sensors. (Note that the ID will have a "G" or "Y" which is for "green" and "yellow" to aid in cart identification in the software). Select position for the channel for each cart.

(8) Set up the View for two graphs: velocity vs. time for both carts. Make the green cart the top graph. If you prefer, you can use one graph, with time on the x-axis and both velocities on the y-axis (they will be in different colors).

(9) Practice creating a gentle collision: place Cart 2 at rest in the middle of the track. Push Cart 1 so that it rolls toward Cart 2 with the hook-and-pile pads toward one another. The carts should collide, stick together, and then continue to roll together.

(10) Click Collect to start data collection. Repeat the collision you practiced and use the velocity graphs to verify that the carts are recording throughout the entire range of motion.

Part I – Hook and Pile Bumpers

(11) Return Cart 2 to the middle of the track and Cart 1 to its starting position. Click Collect to begin taking data and repeat the collision.

(12) From the velocity graphs, you can measure the mean velocity before and after the collision for each cart. Highlight a time interval, click Graph Tools, and choose View Statistics. Measure the mean velocity for each cart, before and after the collision, and enter the values in Table 2. Use a small width region just before the collision and just after the collision to minimize error due to frictional losses. [NOTE – the green cart is in blue-green and the yellow cart is in the yellow-green.]

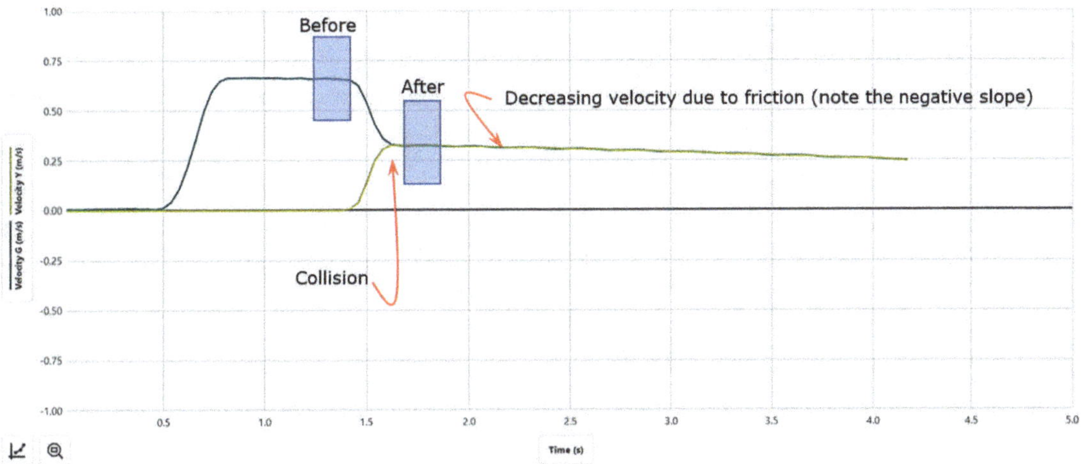

(13) Repeat the process for a second trial.

Part II – Hook and Pile to Empty Bumpers

(14) Remove the hook-and-pile inserts from the yellow cart. Remeasure the mass of this cart and record it in Table 1. The green cart's mass should stay the same.

(15) Face the hook-and-pile bumper on the green cart to the empty bumper on the yellow cart. Practice the collision, starting with the yellow at rest. The carts will not stick together, nor will they bounce apart smoothly.

(16) Return Cart 2 to the middle of the track and Cart 1 to its starting position. Click Collect to begin taking data and repeat the collision. Use the previous procedure to measure and record the cart velocities in Table 2.

(17) Repeat the process for a second trial.

Part III – Magnetic Bumpers

(18) Insert magnets into both carts so that the carts will repel (face the "N" outward on both carts). Measure the masses of the carts and record in Table 1.

(19) Place the carts on the track with the magnetic bumpers facing each other. Practice making this new gentle collision, again starting with the yellow starting at rest. The carts should smoothly repel each other without physically touching.

(20) Return Cart 2 to the middle of the track and Cart 1 to its starting position. Click Collect to begin taking data and repeat the collision. Use the previous procedure to measure and record the cart velocities in Table 2.

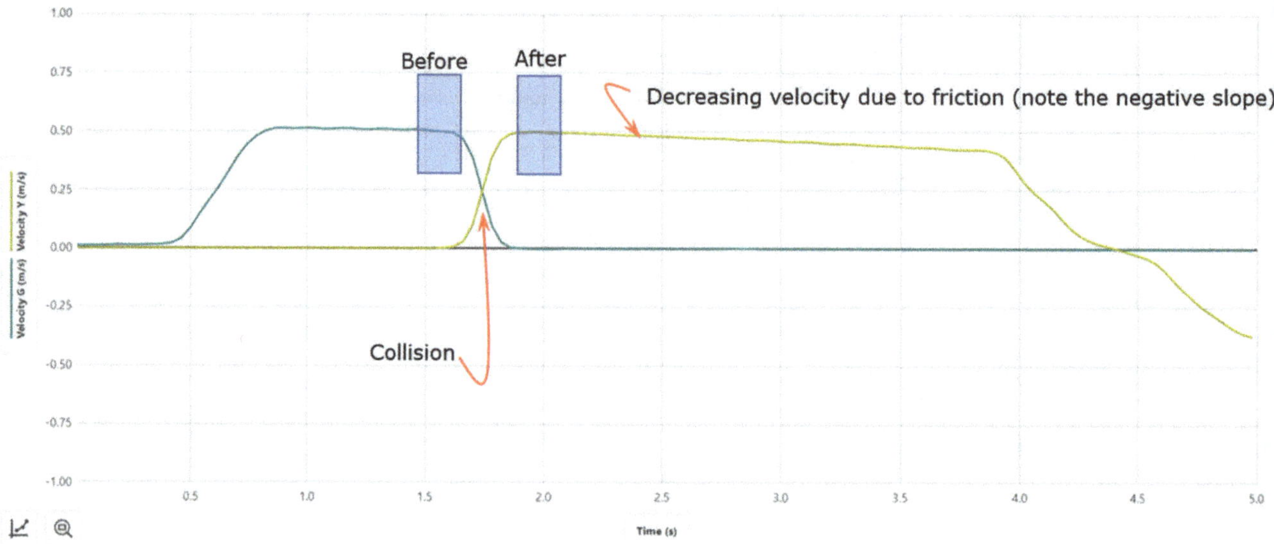

(21) Repeat the process for a second trial.

(22) Complete Questions 1–6.

IMPULSE & MOMENTUM

BACKGROUND

In this part of the lab, we will measure a cart's change in momentum and compare it to the impulse it receives. The impulse-momentum theorem relates impulse, the average force applied to an object times the length of time the force is applied, and the change in momentum of the object:

$$\overline{\mathbf{F}}\Delta t = m\mathbf{v}_f - m\mathbf{v}_0$$

We will use a cart equipped with a hoop spring. Its momentum will change as it collides with the end stop at the end of the track. The hoop will compress and apply an increasing force until the cart stops. The cart then changes direction, as the hoop expands back to its original shape. The force applied by the spring and cart velocity throughout the motion are measured by the sensor cart.

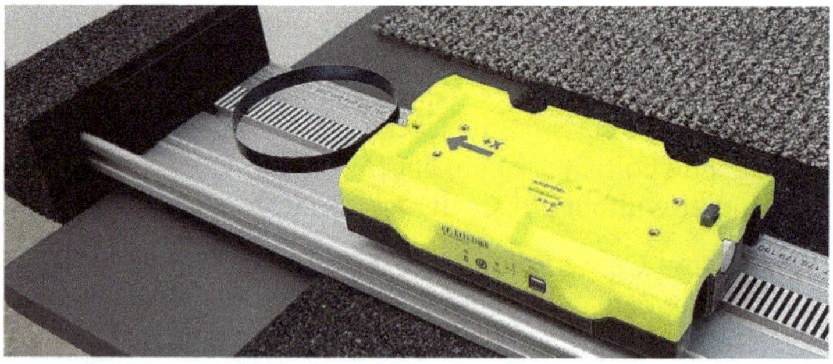

This is an example of an elastic collision. A perfectly elastic collision is defined as one in which there is no loss of kinetic energy in the collision, while an inelastic collision is one in which part of the kinetic energy is changed to some other form of energy in the collision. Any macroscopic collision between objects will convert some of the kinetic energy into other forms of energy, so no large-scale impacts are perfectly elastic. Momentum is conserved in inelastic collisions, but one cannot track the kinetic energy through the collision since some of it is converted to other forms.

PROCEDURE

(1) Attach the hoop spring to the cart. Measure the mass of the cart (with the hoop spring installed) and record the value in Table 5.

(2) Start Graphical Analysis.

(3) Click Sensor Data Collection and select your sensor from the list of Discovered Wireless Devices.

(4) Enable the Position and the Force channels. Click Done.

(5) Reverse the Force Sensor.

(6) Configure the View for two graphs: velocity vs. time, and force vs. time.

(7) Click Mode to enter data collection settings. Change the Rate to 250 samples/second and End Collection to 5 s. Click Done.

Elastic Collisions

(8) Practice releasing the cart so it rolls toward the end stop, bounces gently, and returns to your hand.

(9) Position the cart so that it is about 50 cm from the end stop. Click Collect to start data collection, then roll the cart as you practiced in the previous step.

(10) Study your graphs to determine if the run was useful. Confirm that you can see a region of constant velocity before and after the impact. If necessary, repeat data collection.

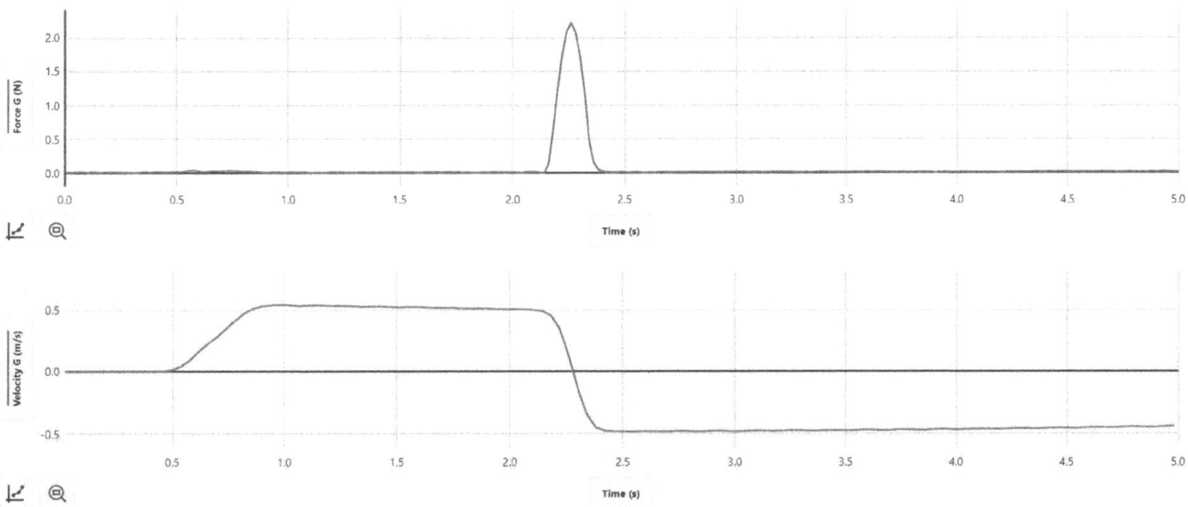

(11) Once you have made a run with good velocity and force graphs, analyze your data. On the Velocity vs. Time graph, highlight an interval just before the impulse, when the cart was moving at approximately constant speed toward the end stop. As before, use a short interval to minimize the effect of friction. Click Graph Tools and choose View Statistics. Record the average velocity before the collision (v_i) in Table 6.

(12) Repeat to determine the average velocity just after the impulse (v_f), when the cart was moving at approximately constant speed away from the end stop. Record this in Table 6 (it will be a negative value to indicate velocity in the opposite direction but record it as a positive value).

(13) To get the impulse, we need to calculate the integral on the force vs. time graph. Highlight the region that represents the impulse (begin at the point where the force becomes non-zero). Click Graph Tools and Select View Integral. Record the value of the integral as Impulse in Table 6.

(14) Repeat the experiment two more times to collect a total of three trials.

Inelastic Collisions

(15) Remove the hoop from the cart. Replace it with the small rubber bumper. Measure the mass of the cart and record the data in Table 5.

(16) Press a golf-ball size mass of Play-Doh clay to the end stop. Use the cart and bumper to form a depression in the clay. The idea is for the bumper to fit into the clay exactly so the cart will come to a stop without rebounding.

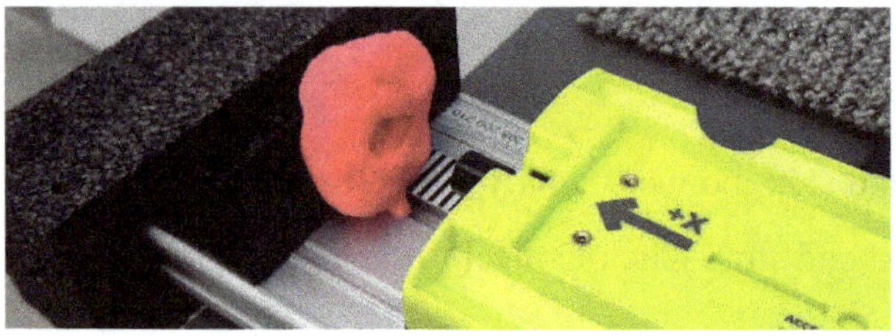

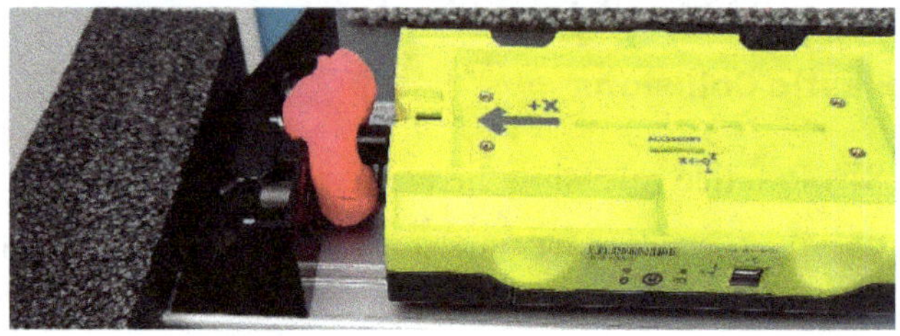

(17) Position the cart so that the cart is approximately 50 cm from the end stop. Click Collect to start data collection, then roll the cart towards the end stop with the clay.

(18) Study your graphs to determine if the run was useful. Confirm that you can see a region of constant velocity before and after the impact. If necessary, reshape the clay and repeat data collection.

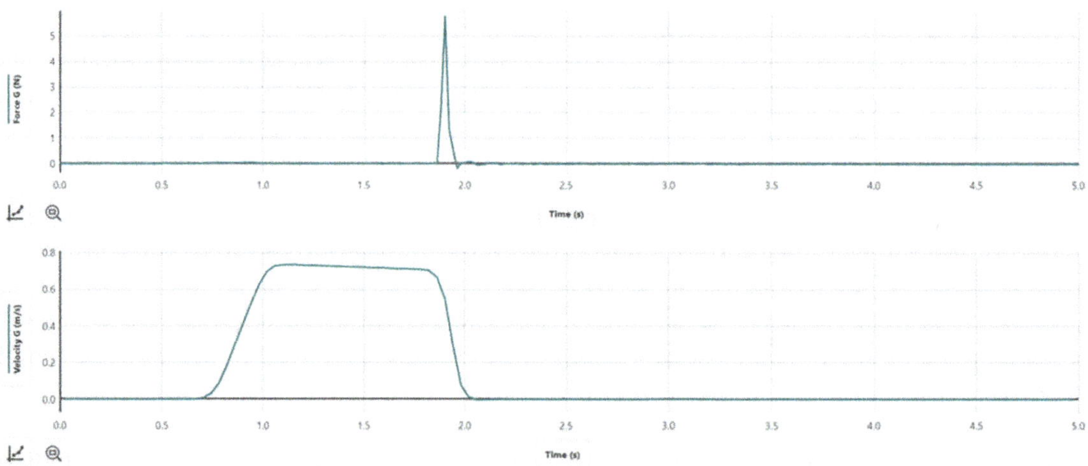

(19) On the Velocity vs. Time graph, highlight an interval just before the impulse, when the cart was moving at approximately constant speed toward the end stop. As before, use a short interval to minimize the effect of friction. Click Graph Tools and choose View Statistics. Record the average velocity before the collision (v_i) in Table 6.

(20) On the Force vs. Time graph, highlight the region that represents the impulse. Click Graph Tools and Select View Integral. Record the value of the integral as Impulse in Table 6.

(21) Repeat the experiment two more times to collect a total of three trials.

(22) Complete Questions 7–10.

LAB 8 – ASSIGNMENT

QUESTIONS – MOMENTUM, ENERGY & COLLISIONS

Table 1

	Part I	Part II	Part III
Mass (kg) of Cart 1 (Green)			
Mass (kg) of Cart 2 (Yellow)			

Table 2

		Velocity (m/s)			
Bumper Type	**Trial**	**Cart 1 (before)**	**Cart 2 (before)**	**Cart 1 (after)**	**Cart 2 (after)**
Hook and Pile	**1**		0		
	2		0		
Mixed	**3**		0		
	4		0		
Magnetic	**5**		0		
	6		0		

(1) Use the data in Tables 1 and 2 to calculate momentum ($\rho = mv$). Place answers in Table 3.

Table 3

	Momentum Before (kg·m/s)			Momentum After (kg·m/s)			
Trial	**Cart 1**	**Cart 2**	**Total**	**Cart 1**	**Cart 2**	**Total**	**Ratio ***
1		0					
2		0					
3		0					
4		0					
5		0					
6		0					

* Total Momentum After / Total Momentum Before

(2) The total momentum ratio (after / before) is < 1.00, which would suggest a slight loss of momentum (i.e., total momentum was not conserved). Why is this the case?

(3) Use the data in Tables 1 and 2 to calculate kinetic energy (½mv^2). Place answers in Table 4.

Table 4

	Kinetic Energy Before (J)			Kinetic Energy After (J)			
Trial	**Cart 1**	**Cart 2**	**Total**	**Cart 1**	**Cart 2**	**Total**	**Ratio ***
1		0					
2		0					
3		0					
4		0					
5		0					
6		0					

(4) Is KE energy conserved in hook and pile collision (Trials 1 & 2)? How about the mixed collision (trials 3 & 4) and the magnetic bumper collisions? How do you know?

(5) Classify the three collision types as elastic, inelastic, or completely inelastic.

(6) A billiard ball moving at 3 m/s strikes a stationary ball of the same mass. After the collision, the first ball moves at 2.33 m/s at an angle of 30° with respect to the original line of motion. Assuming an elastic collision (and ignoring friction and rotational motion), find the struck ball's velocity.

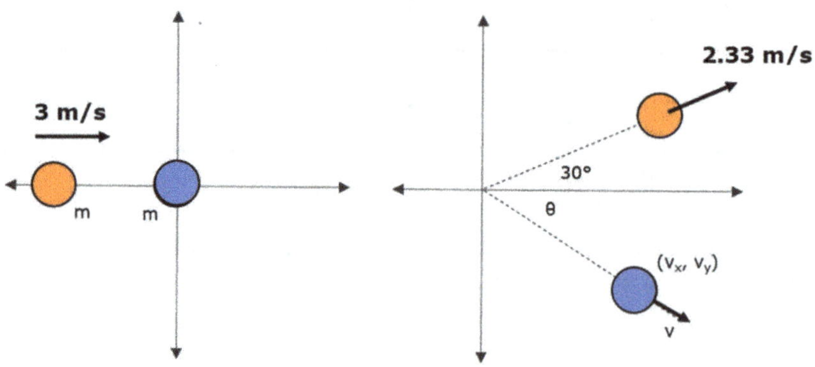

QUESTIONS – IMPULSE & MOMENTUM

Table 5

Mass of Cart + Hoop Bumper (kg)	Mass of Cart + Rubber Bumper (kg)

(7) Complete Table 6 by calculating the change in velocity, the change in momentum, and the ratio of Impulse / Δρ. Calculate the change in velocity by *adding* v_f and v_i (recall one of them was negative).

Table 6

	Velocity (m/s)			Impulse/Change in Momentum		
Trial	**Final (v_f)**	**Initial (v_i)**	**Change (Δv)**	**Δp (N·s) ***	**Impulse (N·s)**	**Ratio**
Elastic 1						
2						
3						
Inelastic 1	0					
2	0					
3	0					

* $\Delta p = m\Delta v$

(8) According to the impulse-momentum theorem, impulse should be equal to the change in momentum. It is likely that your data shows that impulse is slightly lower than the change in momentum. Why?

(9) Mathematically, the impulse-momentum theorem claims: $Ft = \Delta mv \Rightarrow F = \frac{\Delta mv}{t}$. Extending the period over which a force acts, is a common practice to lessen force. For example, the front of a car is designed to crumple in an accident; this increases the time the car takes to stop, which reduces the force. If we want to reduce F by a factor of 1000, how much do we have to increase the time by?

(10) A 0.15 kg baseball is thrown horizontally at 35 m/s. After it is struck by a bat, it is travelling at –35 m/s. Calculate the acceleration of the ball during the time it was in contact with the ball (0.0005 s).

LAB 9 | TORQUE & EQUILIBRIUM

GETTING STARTED

OVERVIEW

In this lab, you will analyze equilibrium forces when torque is involved.

EQUIPMENT

- Lab Computer
- Rubber Mat
- Rod Stand Base
- Rod (Short)
- Aluminum Meter Stick
- Torque Mass Hangers (5)
- Pivot
- Set of Masses
- Force and Acceleration Sensor

TORQUE

BACKGROUND

Torque is a measure of how much a force acting on an object causes that object to rotate. It is the product of the force applied and the lever arm (the perpendicular distance from the axis of rotation to the line of action of the force). Rotation is about the pivot point. Mathematically, torque is:

$$\tau = \mathbf{r}\left|\mathbf{F}\right|\sin(\theta)$$

PROCEDURE

(1) Complete Questions 1–5.

EQUILIBRIUM & TORQUE

BACKGROUND

Equilibrium requires the net forces to be equal. For an object to be in equilibrium, two conditions must be met:

(a) the net force acting on the object must be zero: $\sum F_x = 0,\ \sum F_y = 0$ and

(b) the net torque acting on the object must be zero: $\sum \tau = 0$

These conditions apply for static equilibrium (object is not moving) or dynamic equilibrium (object is moving at a constant velocity). Torque is the rotational equivalent of a force and equilibrium is maintained if the rotational forces are balanced (recall that CCW rotation is positive and CW rotation is negative by definition).

From experience, we know that the same force will be much more effective at rotating an object such as a nut or a door if our hand is not too close to the pivot point. This is why we have long-handled wrenches, and why doorknobs are not next to the door hinges. We can use a simple balance to experiment with creating a system that is in rotational equilibrium. In this diagram, masses m_2 and m_3 will create a negative rotation (CW) and mass m_1 will create a positive rotation (CCW). If these rotational forces are balanced, then the device will be balanced as shown.

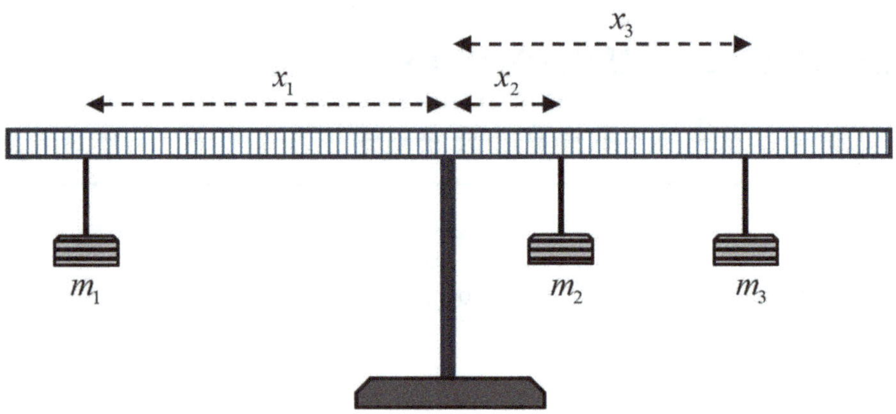

For equilibrium: $\sum \tau = 0 \quad \Rightarrow \quad \sum \tau_{\text{CCW}} = \sum \tau_{\text{CW}} \quad \Rightarrow \quad m_1 x_1 = m_2 x_2 + m_3 x_3$

PROCEDURE

(1) Set up the rod stand on the rubber mat.

(2) Attach the pivot to the meter stick clamp. The rod on the pivot inserts into the top hole in the meter stick clamp (closest to the bubble). You may have to rotate the clamp to find the right orientation.

(3) Attach the pivot to the top of the rod so that the meter stick clamp is on the side. Insert the meter stick into the clamp so that the bubble level is on top. Gently tighten the screw so that the pointer is at the 50 cm mark. Adjust until the bubble level shows that the meter stick is horizontal.

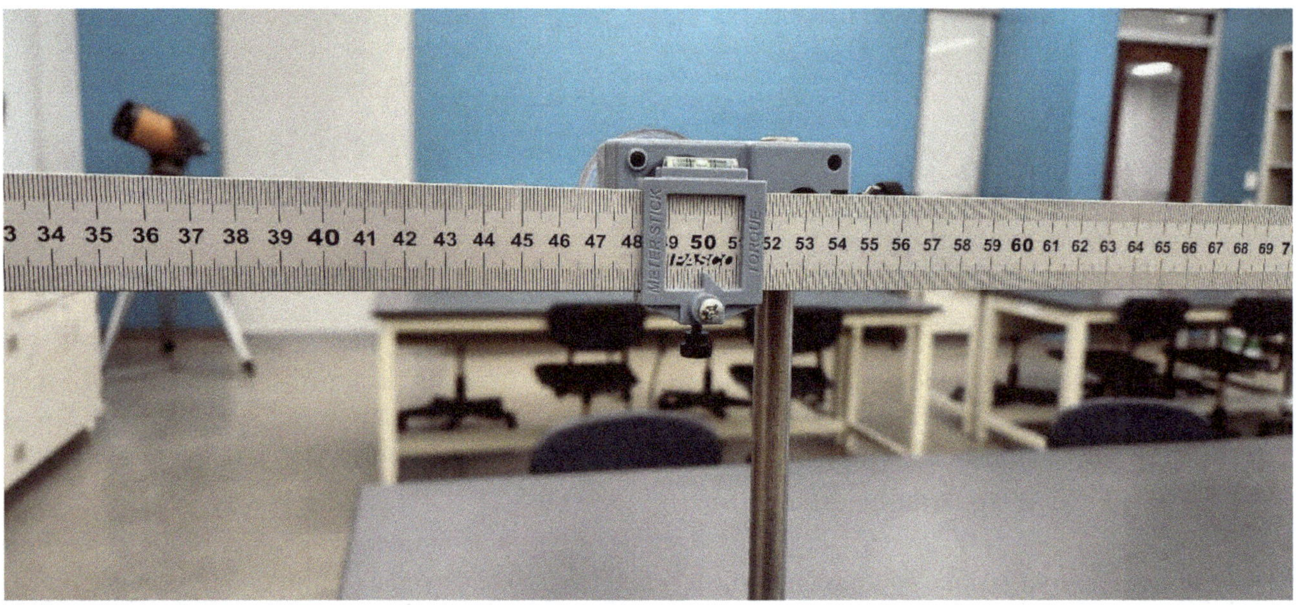

(4) Attach one of the mass hangers to the meter stick so that the screw is on top and the pointer is at the 10 cm mark. This will cause the meter stick to rotate because there is a torque from the gravitational force on the clamp.

(5) Attach a second mass hanger at the 90 cm mark. Ensure that the meter stick balances.

(6) Place a 100 g mass on each hanger. Ensure that the meter stick balances.

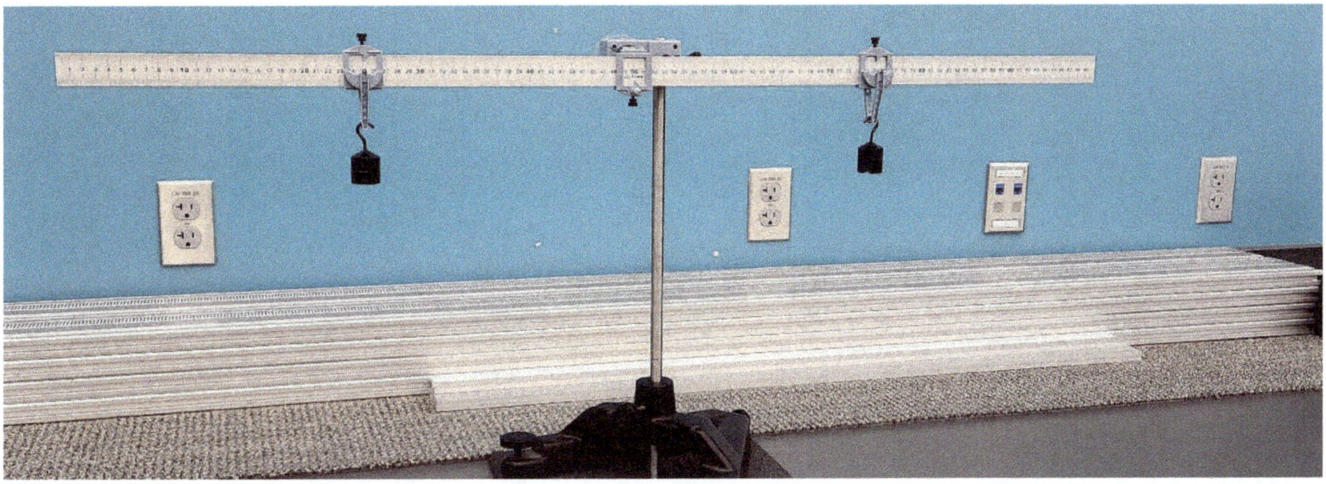

(7) Use the mass hangers and masses to complete Table 1.

Torque with a Force Sensor

(8) Turn on a force sensor.

(9) Start Graphical Analysis.

(10) Click Sensor Data Collection and select your sensor from the list of Discovered Wireless Devices.

(11) Configure the view for one graph (Force vs. time).

(12) Change the data collection parameters to End Collection at 10 s.

(13) Use a mass hanger to place a 200 g mass at 80 cm (this will create a CW torque).

(14) Place a mass hanger at 5 cm.

(15) Zero the force sensor.

(16) Click Collect to begin data collection. Hook the force sensor to the mass hanger and pull down until the meter stick is level. Hold this position until the end of the 10 s.

(17) Highlight the region on the graph corresponding to when the system was level.

(18) Use Graph Tools – View Statistics to see the mean. Record this value in Table 3.

(19) Repeat for the distances indicated in Table 3.

(20) Complete Questions 6–12.

LAB 9 – ASSIGNMENT

QUESTIONS – TORQUE

Use the diagram on the right to answer Q1–Q3. Recall that torque is directional.

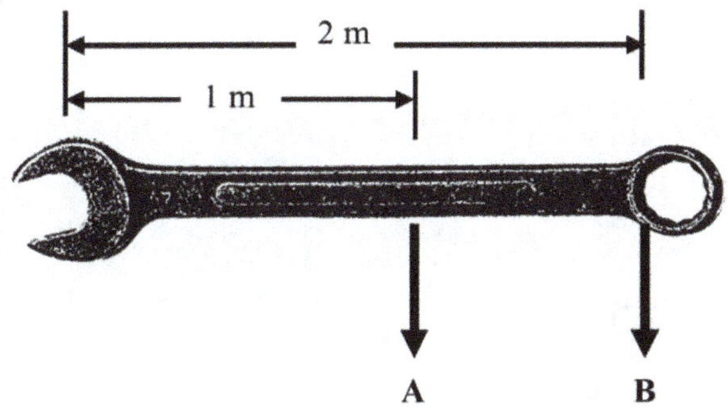

(1) What is the torque on the bolt if the worker pulls straight down at position A with a force of 30 N?

(2) What is the torque on the bolt if the worker pushes straight up at position B with a force of 30 N?

(3) What is the torque on the bolt if the worker pushes up at position A and position B with 20 N each?

(4) While painting a wall, a 55.6 kg painter is standing on a 3.15 m long board that is resting on a pair of sawhorses. The board is homogeneous with a mass of 14.5 kg. The sawhorse on the right is 1.00 m from the right end. How far away can the painter walk from the sawhorse on the right until the board begins to tip?

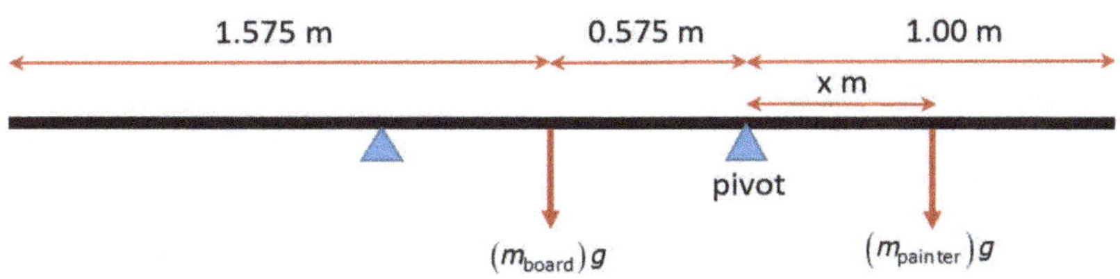

(5) Consider the human arm holding a book. Let's see how changing the mass of the book and the angle of the arm (mass = 2.5 kg) changes the force that the biceps must exert. Compute the force of the biceps (F_B) as shown ($\theta = 90°$ between the upper and lower arm) with a mass of 4 kg, 6 kg, and 8 kg. Then, hold the book's mass at 4 kg and change the angle to 60° and 120°. Show ALL steps.

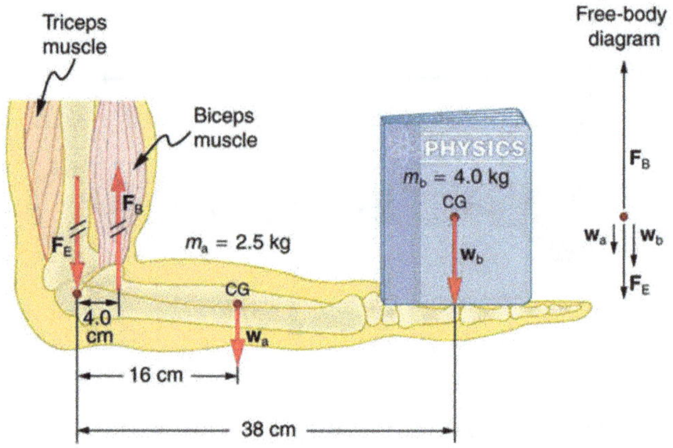

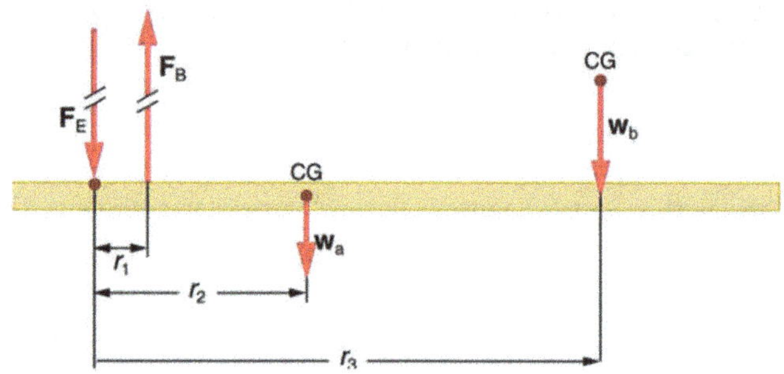

Calculations (changing book mass)

Calculations (changing angle)

QUESTIONS – EQUILIBRIUM & TORQUE

(6) Fill in the blanks in the table. Remember that distances are measured from 0.5 m. The hangers have a mass of 10 g, which needs to be added into the masses in the table).

Table 1

Trial	r_1 (m)	m_1 (kg)	r_2 (m)	m_2 (kg)	r_3 (m)	m_3 (kg)
1	0.400	0.210		0.510		
2	0.200	0.210		0.110		
3	0.200	0.210	0.200	0.110	0.333	0.06
4	0.250	0.210	0.220		0.470	
5	0.150	0.510		0.210		0.210
6	0.150	0.510		0.110		0.060

(7) Use the data in Table 1 to fill in Table 2. Recall torque ($\tau = rF = rmg$) has a sign (+ for CCW, - for CW).

Table 2

	CCW Torque	CW Torque			|Difference|
Trial	τ_1 (N·m)	τ_2 (N·m)	τ_3 (N·m)	Total ($\tau_2 + \tau_3$)	|CCW – CW|
1					
2					
3					
4					
5					
6					

(8) Fill in the table based on your data. Note that the mass hanger at 5 cm is 0.45 m from the pivot (r_1).

Table 3

CCW Torque (Force Sensor)			CW Torque (Hanging Mass)			|Difference|
r_1 (m)	F (N)	τ_1 (N·m)	r_2 (m)	m_2 (kg)	τ_2 (N·m)	|CCW – CW|
0.45			0.30	0.210		
0.40			0.30	0.210		
0.35			0.30	0.210		
0.30			0.30	0.210		
0.25			0.30	0.210		
0.20			0.30	0.210		
0.15			0.30	0.210		
0.10			0.30	0.210		

(9) Use Graphical Analysis in the manual mode to plot r_1 (distance from the pivot to the force sensor) on the x-axis against Force on the y-axis. Find the best fitting curve (looking for lowest RMSE). Report the curve family (power, quadratic, etc). and the specific equation (e.g., $3x^{1.2}$ rather than ax^b). Also, report the RMSE.

(10) Consider the system shown (a simple version of a child's mobile). If A = 12 kg, what is the mass of C?

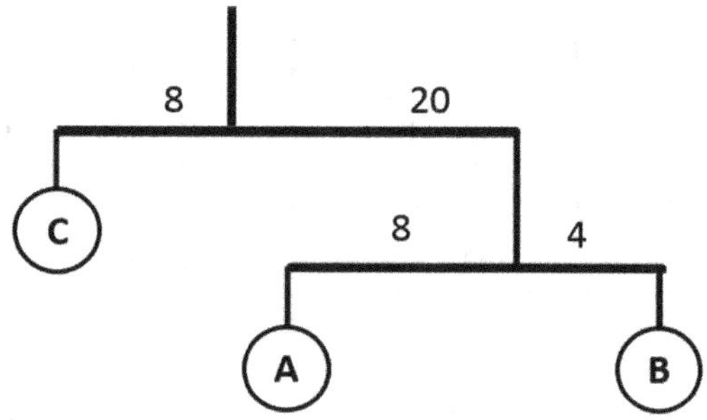

(11) Now for a more complicated mobile. We're to build a mobile of a solar system that is in equilibrium as shown in the diagram. Find the mass of each planet, B through H, given that A has a weight of 5 N. [HINT – solve them in order B to H].

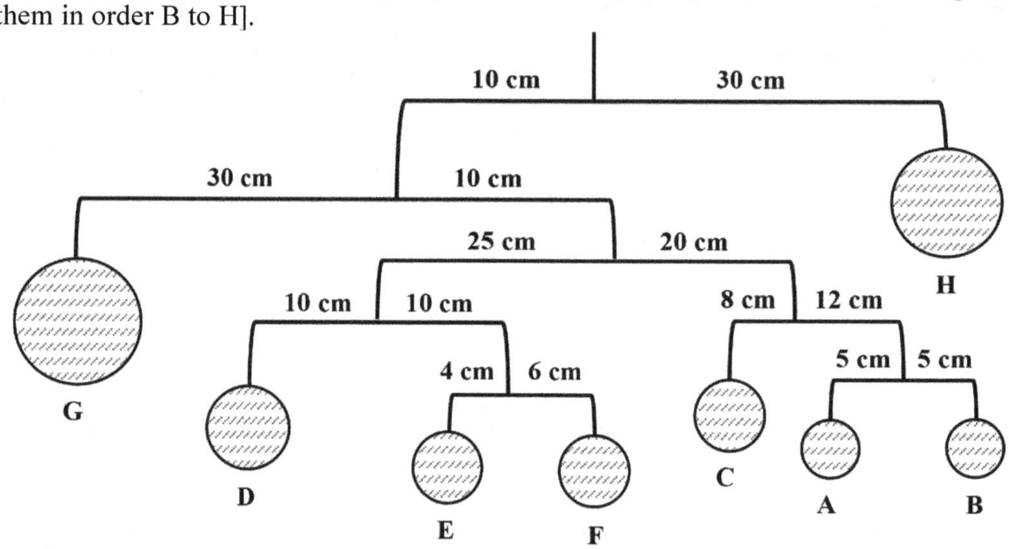

LAB 10 | THERMAL EXPANSION

GETTING STARTED

OVERVIEW

In this lab, you will experimentally determine the coefficient of linear expansion for three different metals (brass, copper, and aluminum).

EQUIPMENT

- Thermal Expansion Apparatus
- Steam Generator
- Meter Stick
- Multi-meter
- Patch Cords (2)
- 150 ml Beaker

LINEAR EXPANSION

BACKGROUND

Most materials expand somewhat when heated through a temperature range that does not produce a phase change. The additional heat increases the average amplitude of vibration of the atoms in the material which increases the average separation between the atoms. So, the material expands.

Suppose an object of length L undergoes a temperature change of magnitude ΔT. If ΔT is not overly large, the change in length, ΔL, is proportional to L and ΔT:

$$\Delta L = \alpha L \Delta T$$

where α is the coefficient of linear expansion for a given material. This expansion must be accounted for when building or manufacturing something using expandable materials. Nearly all materials expand when heated; however, the expansion is so small on "human" sized objects that most of the time, expansion can be safely ignored. On very large objects such as bridges, railroads, and large buildings, this is not the case and thermal expansion joints are often incorporated into the design.

NOTES ON TEMPERATURE MEASUREMENT

To make a measurement, a metal tube is placed on the top position of the apparatus and heated by passing steam through it. The tube's expansion is measured with 0.01 mm resolution. Temperature is measured using a 10 kΩ thermistor that converts attached to the center of each metal tube.

A thermistor's resistance varies with temperature. The resistance can be measured with a digital multi-meter and converted to a temperature measurement using the conversion table below. The thermistor used to measure the tube temperature is embedded in that thermistor lug that is fastened to the tube. Once thermal equilibrium has been reached, the heat is highly uniform along the length of the tube. The foam insulation is used to inhibit heat loss through the thermistor lug, so the lug temperature closely follows the tube temperature.

THERMISTOR: TEMPERATURE VERSUS RESISTANCE							
55,142 Ω	-10 °C	26,645 Ω	4 °C	13,676 Ω	18 °C	7,405 Ω	32 °C
52,235	-9	25,357	5	13,068	19	7,100	33
49,499	-8	24,138	6	12,491	20	6,810	34
46,924	-7	22,984	7	11,941	21	6,534	35
44,500	-6	21,892	8	11,418	22	6,271	36
42,215	-5	20,858	9	10,921	23	6,019	37
40,057	-4	19,880	10	10,450	24	5,778	38
38,025	-3	18,953	11	10,000	25	5,549	39
36,107	-2	18,074	12	9,572	26	5,329	40
34,298	-1	17,242	13	9,166	27	5,120	41
32,590	0	16,452	14	8,778	28	4,920	42
30,974	1	15,704	15	8,409	29	4,729	43
29,448	2	14,992	16	8,058	30	4,546	44
28,007	3	14,317	17	7,724	31	4,371	45

THERMISTOR: TEMPERATURE VERSUS RESISTANCE							
4,204 Ω	46 °C	2,489 Ω	60 °C	1,531 Ω	74 °C	975 Ω	88 °C
4,045	47	2,401	61	1,480	75	945	89
3,892	48	2,317	62	1,432	76	917	90
3,745	49	2,236	63	1,386	77	889	91
3,601	50	2,158	64	1,341	78	862	92
3,471	51	2,084	65	1,298	79	837	93
3,342	52	2,012	66	1,256	80	812	94
3,219	53	1,942	67	1,216	81	788	95
3,101	54	1,877	68	1,178	82	765	96
2,988	55	1,814	69	1,141	83	742	97
2,880	56	1,753	70	1,105	84	721	98
2,776	57	1,694	71	1,071	85	700	99
2,677	58	1,637	72	1,038	86	680	100
2,580	59	1,583	73	1,006	87		

PROCEDURE

(1) Assemble the thermal expansion apparatus.

(2) Measure the length of each tube at room temperature (from the inner edge of the larger circular disk on one end to the inner edge of the smaller circular disk at the other end). Record your results in Table 1.

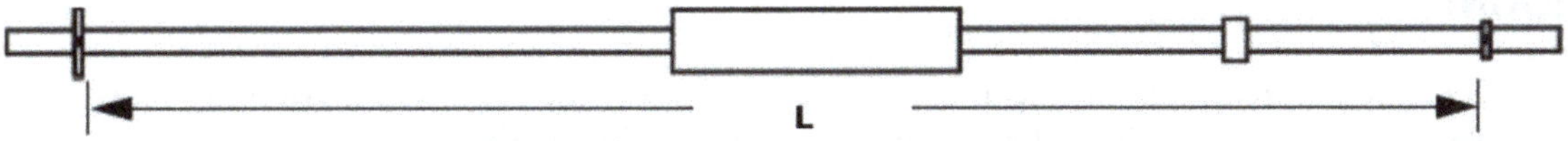

(3) Mount the aluminum tube in the apparatus frame as shown earlier. The smaller circular disk on the tube fits into a slot on the "high" end frame (it will snap into place). The larger circular disk on the tube presses against the tip of the spring arm of the dial indicator.

(4) Turn the metal tube so that the thermistor lug under the foam insulation is on top (you can look for the hole in the top of the large disc – it should be up). Connect the phone plug on the end of the thermistor cable into the phone port on the "high" end frame.

(5) Tighten the thumbscrew in the "high" end frame against the tube until it can no longer be moved and make sure that the foam insulator is centered over the thermistor lug.

(6) Plug the leads of your multi-meter into the banana plug connectors on the "high" end frame just under the phone plug port. Red goes on the bottom. Plug the other end of the connectors into the appropriate ports on the multi-meter.

(7) Turn on the multi-meter by moving the dial to the Ω setting. The meter should auto-scale for us.

(8) Measure R_{room}, the resistance of the thermistor at room temperature. Record this value in Table 1.

(9) Fill the steam generator with ~600-650 mL of water (½ to ¾ full) and place the rubber stopper firmly on the top. Connect the two pieces of rubber tubing to the stopper.

WARNING

- High temperature. The metal tube gets hot during the experiment. Steam can cause burns. Do not touch the metal tube during operation.

(10) Attach the tubing adapter of the rubber tubing to the end of the aluminum tube (attach it to the end farthest from the digital indicator). Clamp the other piece of tubing so that steam cannot escape.

(11) Place a 150 ml beaker under the tube at the "low" end to catch any steam run-off.

(12) Turn the steam generator ON and turn the power dial to maximum (8 or full CW). At this setting it will take 10–15 minutes to boil the water in the chamber and generate steam.

WARNING

- High temperature. The stainless steel tank on the steam generator gets hot when the unit is on.

(13) Press the ON/OFF button on the digital indicator to turn it on. Press the ZERO button to set the initial digital reading to zero. As the tube expands, the spring pin of the digital indicator will stay in contact with the larger circular disk on the tube.

(14) As steam begins to flow, watch the digital display and the resistance reading on the multi-meter. When the thermistor resistance stabilizes, record the resistance (R_{hot}) in Table 1. Also record the expansion of the tube length (ΔL) as indicated by the display on the digital indicator.

(15) Repeat the experiment for the other two metal tubes.

(16) Complete Question 1.

THERMAL EXPANSION & THERMAL STRESS

BACKGROUND

Of course, objects expand in all dimensions (see figure). Therefore, their areas and volumes, as well as their lengths, increase with temperature. The changes in area and in volume are given by:

$$\Delta A = 2\alpha A \Delta T \quad \text{and} \quad \Delta V = \beta V \Delta T$$

where β is the coefficient of volume expansion and $\beta \approx 3\alpha$.

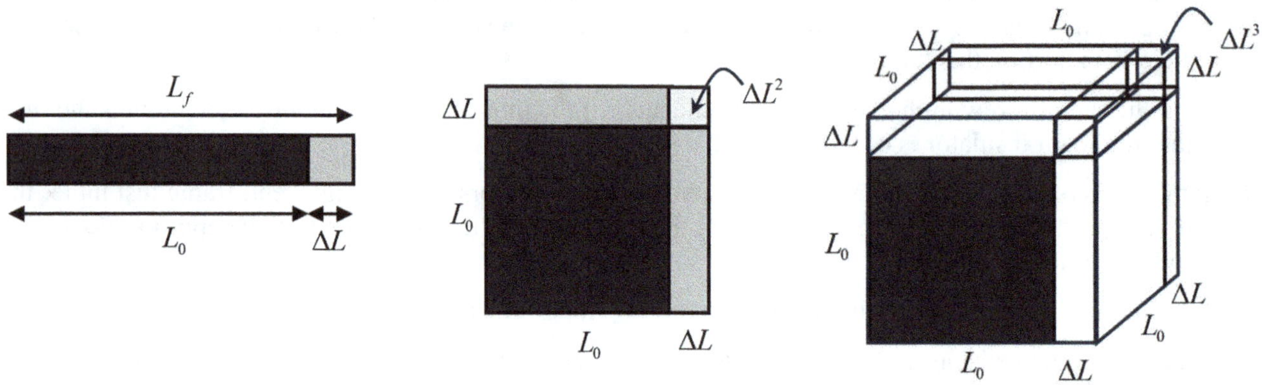

THERMAL STRESS

Thermal stress is created by thermal expansion or contraction. Forces and pressures created by thermal stress can be quite large. For example, railroad tracks and roadways can buckle on hot days if they lack sufficient expansion joints.

The change in volume due to thermal stress (force per unit area) is:

$$\Delta V = \frac{1}{B}\frac{F}{A}V_0$$

where F/A is pressure, V_0 is the original volume and B is the bulk modulus for the material.

PROCEDURE

(1) Complete Questions 2–6.

LAB 10 – ASSIGNMENT

QUESTIONS – LINEAR EXPANSION

Table 1

Tube	Data				Calculations*		
	L (mm)	R_{room} (Ω)	ΔL (mm)	R_{hot} (Ω)	T_{room} (°C)	T_{hot} (°C)	ΔT (°C)
Aluminum							
Brass							
Copper							

*Use the conversion table to convert your thermistor resistance measurements R_{room} and R_{hot} to temperature measurements T_{room} and T_{hot}. You will have to interpolate (assume linear relationship).

(1) Calculate the coefficient of linear expansion (α) for aluminum, brass and copper. Given the accepted values: aluminum ($2.3 \cdot 10^{-5}$/°C), brass ($1.9 \cdot 10^{-5}$/°C), and copper ($1.7 \cdot 10^{-5}$/°C), calculate your percent error.

Table 2

Property	Material		
	Aluminum	Brass	Copper
Coefficient of Linear Expansion (α)			
Percent Error			

Show your work here:

QUESTIONS – THERMAL EXPANSION & THERMAL STRESS

(2) A farmer has a 100 L steel tank on his tractor for fuel. He fills the tank early in the morning when it is 15°C. So, both the steel tank and the fuel are at 15°C. He is distracted and cannot get to the tasks requiring the tractor until later in the day when it's 30°C. The fuel ($\beta = 950\cdot10^{-6}$ /°C) has expanded more than the steel ($\beta = 35\cdot10^{-6}$ /°C) tank due to the temperature change. How much of the gasoline has spilled?

(3) Consider the tractor's 100 L steel tank full of gasoline as discussed in the previous problem. Assume that the amount of gasoline lost due to thermal expansion was 1.1 L (NOT the correct answer to the previous problem). What pressure would be created in the steel tank if the gasoline increases in temperature from 15°C to 30°C without being allowed to expand? The bulk modulus B for fuel is $1\cdot10^9$ N/m^2.

(4) Older thermometers consisted of long thin tubes that were partially filled with mercury. What is the change in length of a 5 cm long column of mercury if its temperature changes from 15°C to 75°C, assuming the mercury ($\alpha = 180\cdot10^{-6}$ /°C) is free to expand linearly?

(5) Surveyors use a steel tape to measure distances. They do their surveying work in all sorts of weather and temperatures. What is the difference in the length of a 30 m long surveyor's tape at -10°C and at 40°C?

(6) If we expand in three dimensions (see below), the math would suggest that $\Delta V \neq 3\alpha\Delta T$ as stated earlier. Explain the discrepancy.

$$V_0 = L_0^{\,3} \quad \text{and} \quad V_f = V_0 + \Delta V \quad \text{and} \quad \Delta L = \alpha L_0 \Delta T \quad \text{then,}$$

$$V_f = \left(L_0 + \Delta L\right)^3 = L_0^{\,3} + 3L_0^{\,2}\Delta L + 3L_0\Delta L^2 + \Delta L^3 = V_0 + \Delta V \quad \Rightarrow$$

$$\Delta V = 3L_0^{\,2}\Delta L + 3L_0\Delta L^2 + \Delta L^3 \quad \text{or} \quad \Delta V = 3L_0^{\,2}\alpha L_0\Delta T + 3L_0\left(\alpha L_0\Delta T\right)^2 + \left(\alpha L_0\Delta T\right)^3$$

LAB 11 | GAS LAWS

GETTING STARTED

OVERVIEW

In this lab, you will experiment with the fundamental gas laws.

EQUIPMENT

- Lab Computer
- Rubber Mat
- Rod Stand Base
- Rod Stand (short)
- Electrode Support
- Gas Pressure Sensor
- Temperature Sensor
- 125 ml Erlenmeyer Flask
- 400 or 600 ml Beaker
- Flask Weight
- 250 ml Beaker (2)
- Transfer Pipette

PRESSURE VS VOLUME

BACKGROUND

In this section, you will determine the relationship between the pressure and volume of a confined gas. The gas we use is air, which will be confined in a syringe connected to a gas pressure sensor. When the volume of the syringe is changed by moving the piston, a change will occur in the pressure exerted by the gas.

Collectively, the fundamental gas laws describe the relationship between pressure, temperature, volume and amount of gas. Technically, the gas laws are applicable to "ideal" gasses, which have several properties:

- an ideal gas consists of a large number of identical molecules
- the volume occupied by the molecules is negligible compared to the volume occupied by the gas
- the molecules obey Newton's laws of motion, and they move in random motion
- the molecules experience forces only during collisions; any collisions are completely elastic, and take a negligible amount of time

Boyle's Law tells us that the volume of gas increases as the pressure decreases, if temperature and the number of molecules are held constant. Mathematically,

$$P_1V_1 = P_2V_2 \quad \text{or} \quad \frac{P_1}{P_2} = \frac{V_2}{V_1}$$

where, P_1 and V_1 are the initial pressure and volume states, and P_2 and V_2 are the pressure and volume of the gas after any change. A plot of pressure versus inverse volume should give a straight line

PROCEDURE

(1) Turn on a gas pressure sensor.

(2) Start Graphical Analysis.

(3) Click Sensor Data Collection and select your sensor from the list of Discovered Wireless Devices.

(4) Configure the view for one graph, one data table and one meter.

(5) In Data Collection Settings, change the mode to Event Based – Events with Entry. Enter volume as the name and mL as the units.

(6) With the syringe **disconnected** from the Gas Pressure Sensor, move the piston of the syringe until the front edge of the inside black ring is positioned at the 5.0 mL mark. See the drawing below. The arrow marks the front edge of the ring – do not base the volume off of the small dome on the plunger.

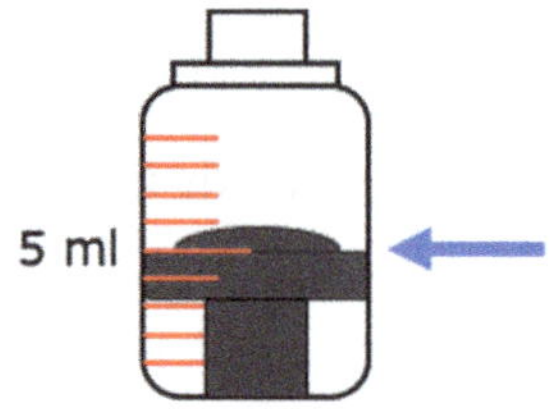

(7) Data collection will be easier if one person manipulates the syringe while another operates the computer.

(8) Connect the syringe to the pressure sensor and click Collect to begin data collection. Click Keep and enter 5 in the volume field. Then click Keep Entry. Also record your entries in Table 1.

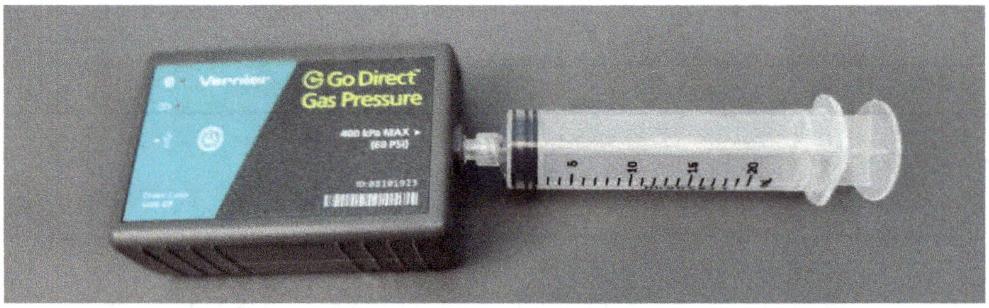

(9) Slowly pull on the piston until the volume is increased to 7.0 mL. When the pressure reading stabilizes, click Keep and enter the volume.

(10) Repeat the process for 9.0, 11.0, 13.0 and 15.0 mL. Record all data in Table 1.

(11) Click Stop to stop data collection, then remove the syringe from the sensor.

(12) In Table 1, there is a row for "Corrected Volume". An improved fit to the data is obtained by correcting for the fact that the recorded volume does not include the small amount of air, about 0.8 mL, inside the sensor box between the end of the syringe and the sensing element. By creating a new Calculated Column and adding 0.8 to the volume reading, you can get a graph of pressure *vs.* corrected volume.

(13) Plot Corrected Volume (x-axis) against Pressure (y-axis) and apply an inverse curve fit.

(14) Complete Questions 1–4.

PRESSURE VS NUMBER OF PARTICLES

BACKGROUND

In this part, you will determine the relationship between the pressure and the number of particles using volume as a proxy. Avogadro's Law states that under the same pressure and temperature conditions, equal volumes of all gases contain the same number of molecules. The number of molecules or atoms in a specific volume of an ideal gas is independent of their size or the molar mass of the gas. Mathematically,

$$\frac{V}{n} = k \quad \text{or} \quad V = nk$$

where, V is volume, n is the number of moles of the gas, and k is a proportionality constant. The constant, k, will depend on the number of moles and the pressure. Alternatively, we can express Avogadro's Law as:

$$\frac{V_1}{n_1} = \frac{V_2}{n_2} \quad \text{or} \quad \frac{V_1}{V_2} = \frac{n_1}{n_2}$$

where V_1 and n_1 are the initial conditions, while V_2 and n_2 are the final conditions.

PROCEDURE

(1) Open a new file in Graphical Analysis.

(2) Change the data collection mode to Events with Entry. Use number as the name and "puffs" as the unit.

(3) Click Start to start data collection.

(4) With the syringe disconnected, position the piston so that the syringe contains one "puff" (2 mL) of air.

(5) Connect the syringe to the sensor and pull back on the piston until the volume reads 10 mL. When the pressure reading stabilizes, click Keep and enter 1 as the number of puffs of air in the syringe. Record your data in Table 2.

(6) Disconnect the syringe from the sensor and move the piston so that it now contains two puffs of air. Reconnect the syringe and move the piston until the volume again reads 10 mL. When the pressure reading stabilizes, click Keep and enter 2 as the number of puffs of air in the syringe.

(7) Repeat the process until you have 5 "puffs" of air. Record all data in Table 2.

(8) Plot number of "puffs" (x-axis) against Pressure (y-axis) and apply a linear curve fit.

(9) Complete Questions 5–8.

PRESSURE VS TEMPERATURE

BACKGROUND

In this section, you will determine the relationship between the pressure and the temperature of a fixed volume of gas. The molecules that make up a gas move in straight lines until they encounter another molecule or the walls of a container. When a molecule encounters a wall, it bounces off and moves in a different direction. When this happens, Newton's Third Law states that the molecule and the wall will experience a force (pressure). Increasing the temperature of a fixed volume of gas causes individual gas molecules to move faster. As the molecules move faster, they encounter the walls of the container more often and with more force. In a rigid container, the more frequent and forceful collisions result in higher pressure. However, if the container is not rigid (e.g., a balloon), the volume will increase, and the pressure will remain the same. Charles' Law is the description of this relationship; it says that the volume of a gas is directly proportional to the temperature, given that the pressure and the amount of gas are constant. Mathematically,

$$\frac{V_1}{T_1} = \frac{V_2}{T_2}$$

where V_1 and T_1 are the initial conditions, while V_2 and T_2 are the final conditions. Note that temperature must be measured in Kelvin (K). We can combine the gas laws into the ideal gas law:

$$PV = nRT \quad \text{or} \quad PV = NkT$$

where P is pressure, V is volume, T is temperature, n is number of moles, N is the number of molecules, R is the gas constant, and k is the Boltzmann constant ($k = 1.38 \cdot 10\text{-}23$ J/K). The value of R depends on the units, usually this is 8.3145 J mol^{-1} K^{-1} or 8.3145 m^3 Pa mol^{-1} K^{-1}.

PROCEDURE

(1) Attach the electrode support to the rod stand and place so that the temperature probe will reach the water bath.

(2) Turn on a Temperature Probe.

(3) In addition to the pressure sensor, connect a Temperature Probe to your interface. Choose New from the File menu. Configure the view for one graph, one data table, and two meters.

(4) Change the data collection mode to Event Based - Selected Events.

(5) Instead of the syringe, you will use an Erlenmeyer flask as your container. Using the stopper and plastic tubing, attach the free end to the pressure sensor. Place the stopper firmly in the flask, and then place the flask in the 400 or 600 mL beaker that serves as the water bath for controlling the gas temperature.

(6) Place the flask weight around the Erlenmeyer flask to keep it from floating. Ensure that the valve is closed.

(7) Click Collect to begin data collection.

(8) Obtain approximately 150 mL of boiling water in one of your spare 250 mL beakers. Pour it into your 400 mL water bath.

(9) Note that the temperature probe is placed in the water bath rather than directly in the gas itself; this means the system must be allowed to equilibrate for several seconds before recording the pressure and temperature. When equilibrium is reached (use the meters to see when the values stabilize), click Keep – Keep Entry; the program will save the pressure and temperature data for each trial. Also record your data in Table 3.

(10) Use the transfer pipette to remove about 20 mL of water from the water bath, then add a small amount of ice to lower the temperature by about 10–15°C. You will not need much ice.

(11) When the temperature and pressure readings stabilize, click Keep – Keep Entry.

(12) Repeat several more times, reducing the temperature by 10–15°C each time. Record all data in Table 3 (note that you may not need all columns in Table 3). You want your final temperature to be approximately 5°C.

(13) Stop data collection. Plot temperature (x-axis) against pressure (y-axis) and apply a linear curve fit.

(14) Complete Questions 9–13.

LAB 11 – ASSIGNMENT

QUESTIONS – PRESSURE VS VOLUME

Table 1

Trial	1	2	3	4	5	6
Volume (ml)						
Corrected Vol (ml)						
Pressure (kPa)						

(1) Sketch your *P*-*V* curve.

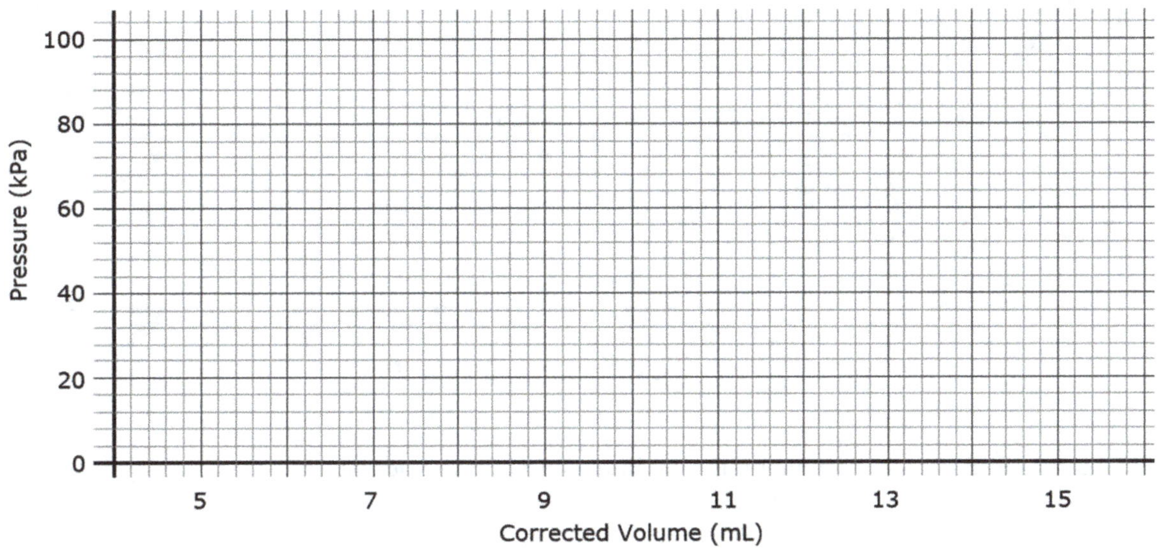

(2) What was the formula and RMSE for the inverse curve fit to your data?

(3) Using your equation in (1), what happens to pressure if we double the volume (you'll have to pick arbitrary volume values). Support your answer with numbers.

(4) Using your equation in (1), what happens to pressure if the volume is halved. Support your answer with numbers.

QUESTIONS – PRESSURE VS NUMBER OF PARTICLES

Table 2

Trial	**1**	**2**	**3**	**4**	**5**
"Puffs"					
Pressure (kPa)					

(5) Sketch your *P*-*n* curve.

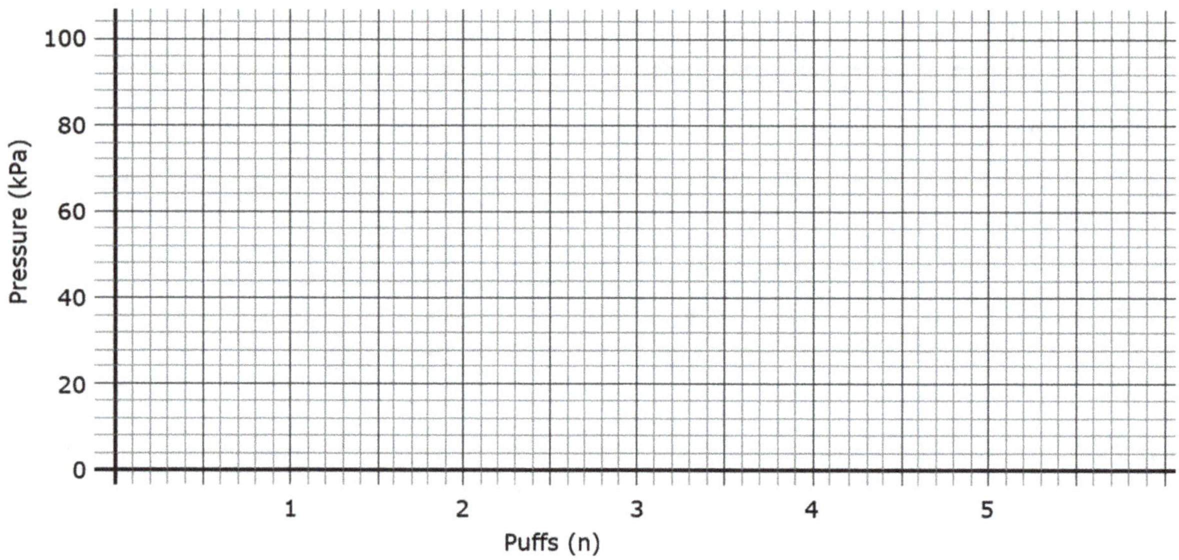

(6) What was the formula and correlation for the linear curve fit to your data.

(7) What does the pressure at 5 puffs of air represent?

(8) Using your equation in (6), what would the pressure be in a vacuum (no air)?

QUESTIONS - PRESSURE VS TEMPERATURE

Table 3

Trial	1	2	3	4	5	6	7	8
Temperature (°C)								
Pressure (kPa)								

(9) Sketch your P-T curve.

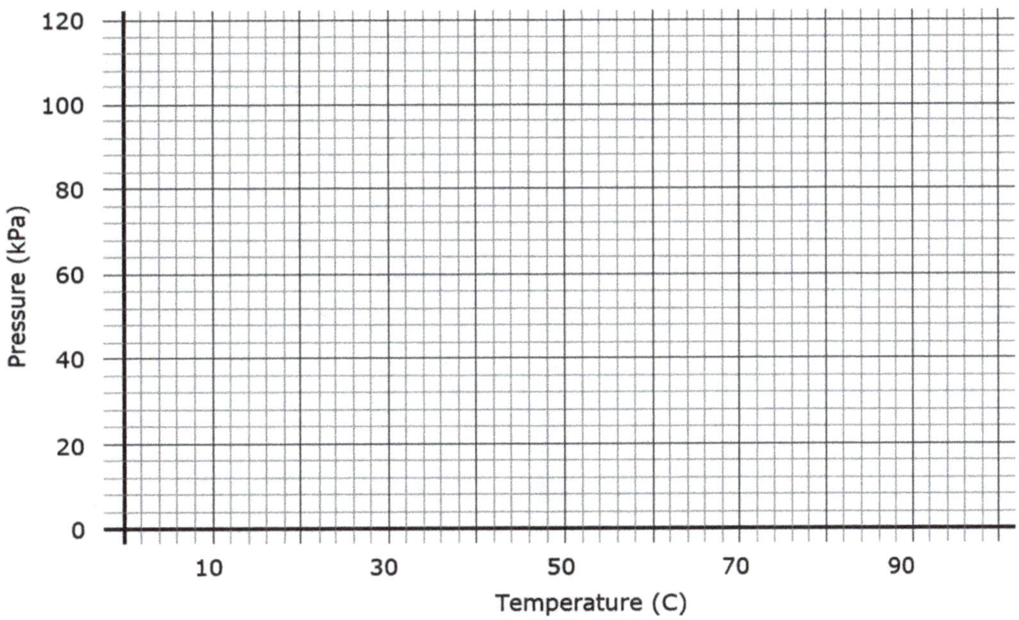

(10) What was the formula and correlation for the linear curve fit to your data.

(11) Use your equation in (10) to predict the temperature at which gas pressure drops to zero. Theoretically, what should this temperature be?

(12) Do a linear fit to your data for 30°C and colder. Report the formula and correlation. Use this equation to predict the temperature at which gas pressure drops to zero. Did this improve (11)?

(13) Airline passengers would find it most comfortable if the cabin air was at a pressure of one atmosphere (101.325 kPa); however, there are engineering and economic reasons to maintain a lower pressure. During a flight the cabin pressure was kept at a compromise value of 81 kPa — much higher than the pressure outside, but still lower than what most people live under (80 kPa is the average atmospheric pressure at an elevation of 2000 m). Assume that the temperature stays constant at 23 °C. When the outside air (–60°C and 26 kPa) is drawn into the cabin from outside, to what temperature does it rise after it is compressed?

LAB 12 | FLUID STATICS

GETTING STARTED

OVERVIEW

In this lab, you will investigate the relationship between pressure and depth in fluids, and Archimedes' Principle.

EQUIPMENT

- Lab Computer
- Rubber Mat
- Rod stand Base
- Rod (medium)
- Adjustable Angle Clamp
- Photogate Rod
- Force Sensor
- Gas Pressure Sensor
- String
- Beakers (2)
- Photogate Pendulum Bob Set
- Electrode Holder
- Electrical Tape
- 5-Gallon Bucket
- Meter Stick
- Clamp

PRESSURE VS DEPTH IN A FLUID

BACKGROUND

The pressure exerted by a static fluid depends only upon the depth of the fluid, the density of the fluid, and the acceleration of gravity. The pressure in a static fluid derives from the weight of the fluid and is given by the expression(s):

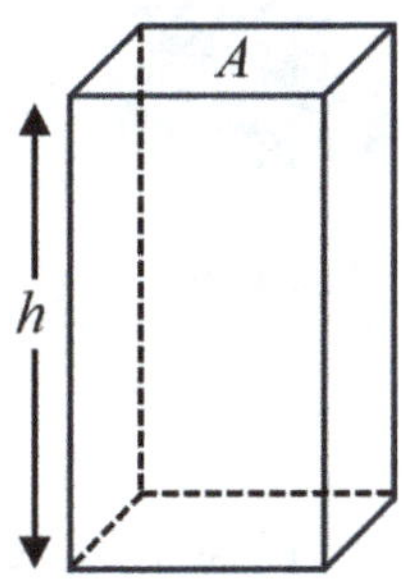

$$P = \frac{F}{A} = \frac{mg}{A} = \frac{\rho V g}{A} = \frac{\rho A h g}{A} = \rho g h$$

P = pressure
F = force
A = area
m = mass
g = gravity
V = volume
h = height of water column (depth)
ρ = density (1000 kg/m^3 for water)

The most remarkable thing about this expression is what it does not include. The fluid pressure at a given depth does not depend upon the total mass or total volume of the liquid.

PROCEDURE

(1) Fill the bucket with water from the sink until it is about half-full. Move the bucket to your work area and put it on the table.

(2) Stand the meter stick along the inside of the 5-gallon bucket. Use a small clamp to secure it to the side of the bucket. Place it so the 0 is at the bottom of the bucket (thus measuring the height of the water level from the bottom of the bucket).

(3) Continue filling the bucket using the beakers until the water depth is exactly 30 cm.

(4) Use the electrical tape to attach the brass pendulum bob to the bottom of the gas pressure tube. This simply adds some weight to help the gas pressure tube to hang somewhat straight.

(5) Assemble the rod base, rod, adjustable angle clamp, and the electrode holder as shown. Note that you will have to disconnect the tubing from the gas pressure sensor to get it through the opening in the electrode support.

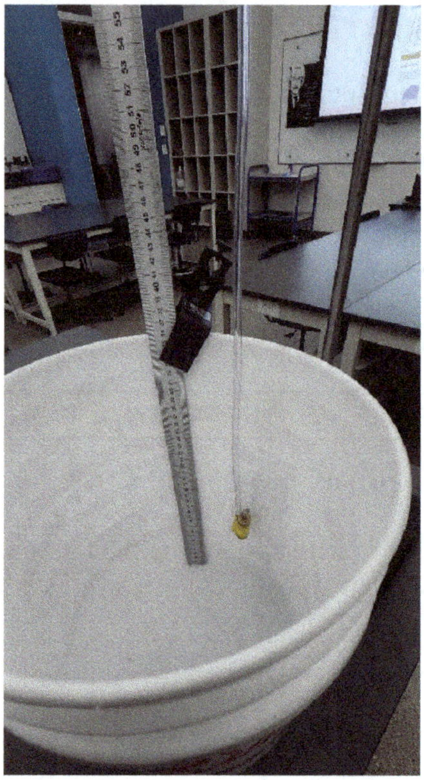

(6) Adjust the height of the electrode support so that the bottom of the gas pressure hose is just <u>above</u> the surface of the water.

(7) Start the Vernier Graphical Analysis application.

(8) Click Sensor Data Collection and select your sensor from the list of Discovered Wireless Devices.

(9) Click Collect to begin data collection, Stop after about 5 seconds. Go to View Statistics and record this pressure in Table 1 under Trial 1. Initial height and new height are just reference points to use to assist in determining the depth. It works best if you use a reference point on the electrode sensor (e.g., bottom of the sensor); it doesn't really matter what you use if you are consistent. Note that the initial and previous heights should be the same for trial 1, which makes the depth 0.

(10) Lower the assembly approximately 2 cm. Note the new height (the initial height will never change).

(11) Click Collect to begin data collection, Stop after about 5 seconds. Go to View Statistics and record the average pressure in Table 1 under Trial 2.

(12) Continue to lower the assembly by approximately 2 cm and repeat the data collection until you just reach the bottom of the bucket (do not set the gas pressure tube on the bottom, but just above it).

(13) Place all data in Table 1.

(14) Restart Vernier Graphical Analysis. Select Manual Entry.

(15) Transfer your data from Table 1 to the data table (x = depth, y = pressure) .

(16) Click Auto Scale, then fit a linear function to the data.

(17) Complete Questions 1–5.

BUOYANT FORCE

BACKGROUND

In this part, you will experimentally determine buoyant force using a force sensor. Archimedes Principle states that the buoyant force on a submerged object is equal to the weight of the fluid that is displaced by the object. Buoyancy results from an unbalanced upward force on the bottom of a submerged object.

The buoyant force is a function of density, volume and gravity:

$$F = \rho g V$$

and can be determined by comparing its mass measured in air with its apparent mass when it is fully submerged in water. When the object is placed underwater, the surrounding fluid exerts an upward buoyant force that reduces the measured weight of the object. The difference between these two measurements corresponds to the mass of the water displaced by the object. Since the buoyant force is equal to the weight of the displaced fluid, this difference can be used to calculate the buoyant force acting on the object.

PROCEDURE

(1) Attach the force sensor to the rod stand. Fill a beaker with water such that the depth is at least 10 cm and place it under the force sensor.

(2) Take a length of string (about 20 cm) and tie a loop at each end. Attach one end of the loop to the force sensor. Raise the force sensor so that the string is well above the water. Attach the brass pendulum bob to the other end of the string, making sure that it is above the water.

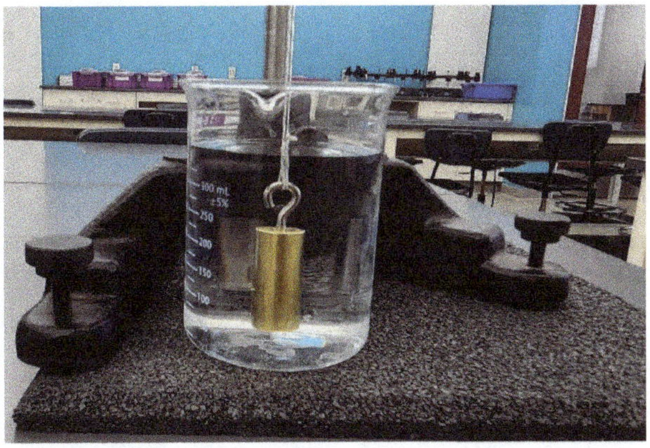

(3) Start the Vernier Graphical Analysis application.

(4) Click Sensor Data Collection and connect the temperature probe. Select your sensor from the list of Discovered Wireless Devices.

(5) With the brass bob hanging in the air, click Collect to begin data collection, Stop after about 5 seconds. Go to View Statistics and record the average force in Table 2.

(6) Lower the bob into the water until its top is fully submerged. Click Collect to begin data collection, stop after about 5 seconds. Go to View Statistics and record the average force.

(7) Raise the bob out of the water. Remove it and dry it off.

(8) Repeat the process with the aluminum and the plastic bob. Record all data in Table 2.

(9) Complete Questions 6–8.

LAB 12 – ASSIGNMENT

QUESTIONS – PRESSURE VS DEPTH IN A FLUID

Table 1

Trial	Initial Height (m)	New Height (m)	Depth (m)*	Mean Pressure (kPa)
1				
2				
3				
4				
5				
6				
7				
8				
9				
10				
11				
12				
13				
14				
15				

*This is the difference between the initial height and the new height.

(1) What was the formula for the linear fit to the data? What was the correlation (r)?

(2) In Q1, what does the y-intercept mean physically?

(3) 1000*slope of your line should be equal the product of the fluid density and the acceleration of gravity. Compute your fluid density by: $1000\left(\frac{\text{slope}}{g}\right)$. Don't forget the units.

(4) Calculate the percent error between your answer in Q3 and the accepted value of 1000 kg/m^3.

(5) A wood salvage company is recovering an old tree trunk from the bottom of a lake. A cable is tied around the log above its center of mass and a force of 9,800 N is applied to the cable to suspend the log in the water (F_{water}), and an additional force of 29,000 N to suspend the log above the lake's surface (F_{air}). What is the volume and the density of the log? Assume a density of 1,007 kg/m^3 for the water.

QUESTIONS – BUOYANT FORCE

Table 2

Material	F_{air} (N)	F_{water} (N)	ΔF (N)
Brass			
Aluminum			
Plastic			

(6) What is your observation regarding ΔF?

(7) Why (regarding your answer to Q6)?

(8) A 75 g block of pine ($\rho = 0.5 \cdot 10^3$ kg/m^3) is held under water by a rope tied to weight on the bottom. If the rope is cut, what will be the acceleration of the block of wood as it ascends?

LAB 13 | FLUID DYNAMICS

GETTING STARTED

OVERVIEW

In this lab, you'll examine the relationship between cross sectional area, pressure, and velocity for fluids in motion.

EQUIPMENT

- Lab Computer
- Venturi Tube
- Short Tube
- Gas Pressure Sensor (2)
- Leaf Blower
- Duct Tape
- Diaper Pail Liner

BERNOULLI'S PRINCIPLE

BACKGROUND

Bernoulli's principle describes the relationship between a fluid's pressure, velocity, and any elevation differences during flow. For fluids with low viscosity, such as water or air, this relationship can be expressed using the Bernoulli equation to estimate how these quantities change along a streamline. We can think of Bernoulli's equation as the conservation of energy as applied to moving fluids. It states that the total energy of the fluid (pressure energy, kinetic energy, and gravitational potential energy) remains constant along a streamline, if energy losses due to friction and viscosity are negligible.

$$P_1 + \frac{1}{2}\rho v_1^2 + \rho g h_1 = P_2 + \frac{1}{2}\rho v_2^2 + \rho g h_2$$

One important consequence of this principle, often referred to as the *Bernoulli effect*, is that regions of faster fluid flow tend to have lower pressure. For example, when a fluid moves through a constricted section of a pipe, its speed increases. Because the total energy of the fluid must remain constant, the increase in kinetic energy is balanced by a decrease in pressure energy. As a result, the pressure in the constricted region is lower than in the wider sections of the flow path

PROCEDURE

(1) Connect two gas pressure sensors to the Venturi tube. You will likely need an extra hose for one of the sensors to get the connections correct.

(2) Insert the short tube into the Venturi Tube flow inlet (shorter side). The tube's diameter is just a bit too small, so you will need to seal this connection with some duct tape.

(3) Start the Vernier Graphical Analysis application.

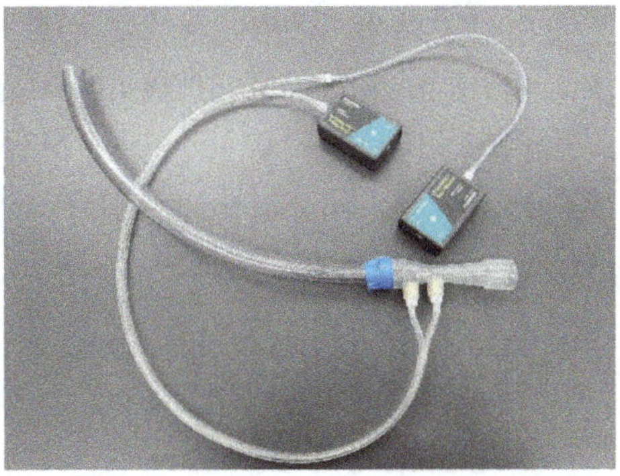

(4) Click Sensor Data Collection and select your sensor from the list of Discovered Wireless Devices.

(5) Go to Data Collection Settings and change the sampling rate to 50 sample/s.

(6) Attach the leaf blower to the short tube and turn it on.

(7) Click Collect to begin data collection.

(8) Click Stop.

(9) Zoom to all data.

(10) Highlight the region where the air flowed through the Venturi tube and get the average pressures via View Statistics. Record the data in Table 1.

(11) Repeat for two additional trials.

(12) Complete Questions 1–2.

ENTRAINMENT

BACKGROUND

Entrainment is a practical application of the Bernoulli Principle. When a fast-moving stream of air or liquid flows past a nearby fluid, the pressure in the fast stream is lower than the surrounding pressure. This pressure difference pulls the nearby fluid into the moving stream, causing it to be carried along with the flow. This process, called entrainment, is used in familiar devices such as gas BBQ grills, chimneys, and paint sprayers.

PROCEDURE

(1) Cut a 1.5 m length of the diaper pail liner and tie one end to seal it. This will form a tube that is closed at one end.

(2) Place the open end of the tube flush against your mouth, take a big breath and blow into the tube. Note how much of the tube a single breath fills up.

(3) Empty the tube of the air. Repeat with a different person.

(4) Empty the tube of air.

(5) Place the open end of the tube about 2-3 inches from your mouth and blow into the tube. Note what happens.

(6) Complete Question 3.

LAB 13 – ASSIGNMENT

QUESTIONS – BERNOULLI'S PRINCIPLE

Table 1

	Air (Pressure in kPa)				Water (Pressure in kPa)			
	Trial 1	Trial 2	Trial 3	Avg	Trial 1	Trial 2	Trial 3	Avg
P_1								
P_2								
ΔP								

You will need the following areas for the questions that follow.

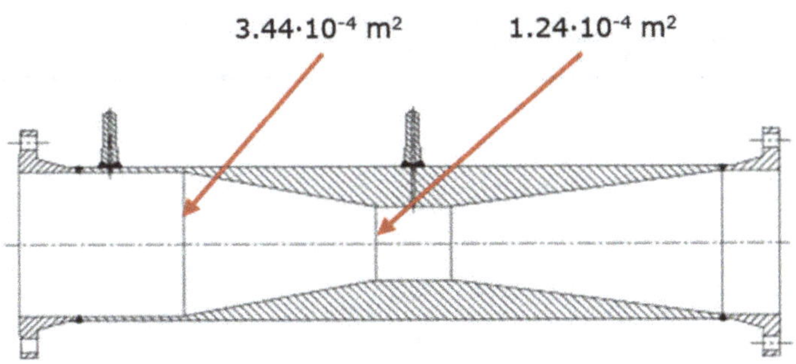

(1) Use Bernoulli's Equation to find the flow rate (v_1 and v_2) in m/s for air in this experiment. Recall that P_1 is the pressure in the inlet area, so v_1 will be the velocity in the inlet area (the larger opening). [**Note** – this is not a trivial calculation.]

(2) Use Bernoulli's Equation to find the flow rate (v_1 and v_2) in m/s for water in this experiment. Recall that P_1 is the pressure in the inlet area, so v_1 will be the velocity in the inlet area (the larger opening).

QUESTIONS – ENTRAINMENT

(3) Explain what happened.

LAB 14 | OPTICS

GETTING STARTED

OVERVIEW

In this lab, you will investigate: (1) reflection and refraction, (2) the relationship between light intensity and distance, and (3) the change in the intensity of light due to cross polarization.

EQUIPMENT

- Lab Computer
- Rubber Mat (2)
- Dynamics Track
- Color and Light Sensor
- Light Sensor Holder
- Light Source
- Polarizer (2)
- Laser Ray Box
- Various lenses

REFLECTION & REFRACTION

BACKGROUND

Light, indeed electromagnetic energy in general, moves in straight lines; however, it can change direction and speed. When light waves meet a boundary with different optical properties, there are several possible results. The light waves can be absorbed, transmitted, scattered, diffracted, reflected or refracted.

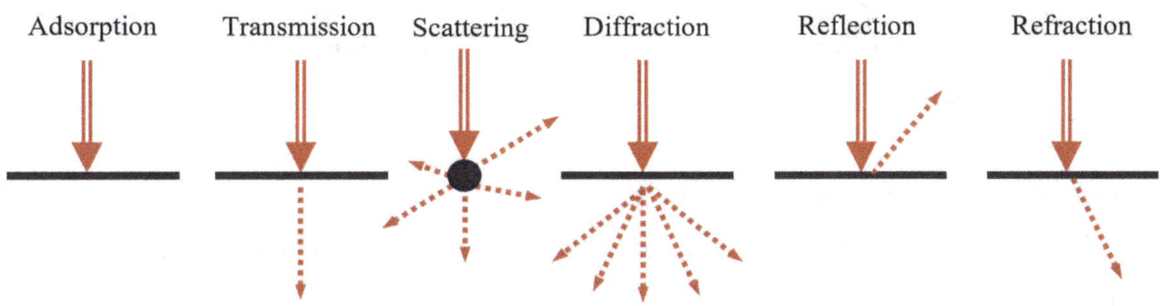

Reflection is the change in the direction of propagation that occurs when light waves are redirected at the boundary between two media. Reflection is governed by the Law of Reflection, which states that the angle of incidence equals the angle of reflection.

$$\theta_i = \theta_r$$

Refraction is the change in speed (and therefore direction) of a wave when it passes from one medium to another. Light waves are refracted because the speed of light is different in different media. Refraction is governed by the Law of Refraction (Snell's Law):

$$n_1 \sin(\theta_1) = n_2 \sin(\theta_2)$$

where n_1 is the index of refraction for material 1, n_2 is the index for material 2, θ_1 is the angle of incidence, and θ_2 is the angle of refraction.

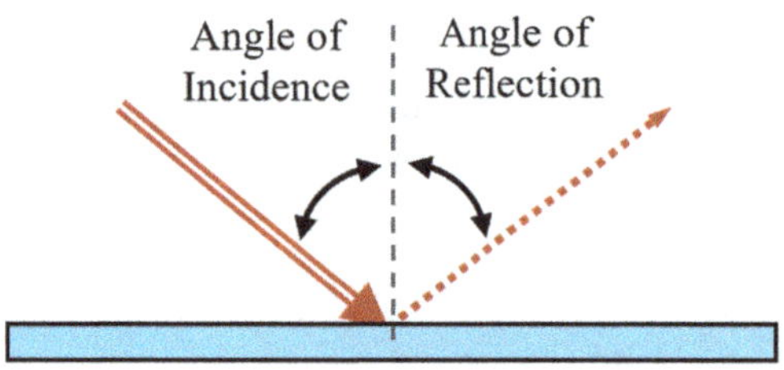

Reflection

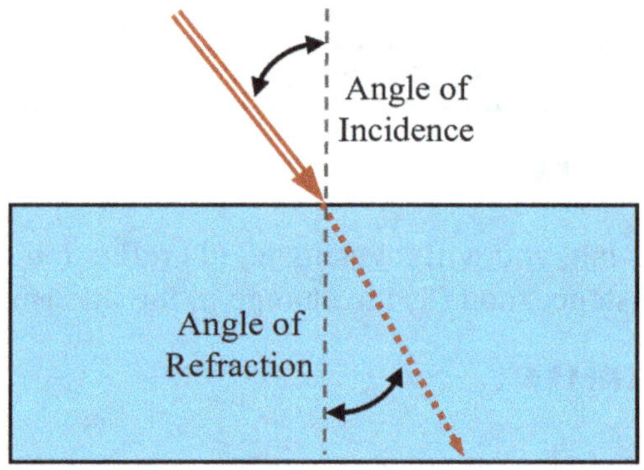

Refraction

A convex or converging lens is shaped so that all light rays that enter parallel to its optical axis intersect at a single point (the focal point) on the opposite side of the lens. Similarly, a concave or diverging lens is shaped so that all rays that enter parallel to its optical axis diverge.

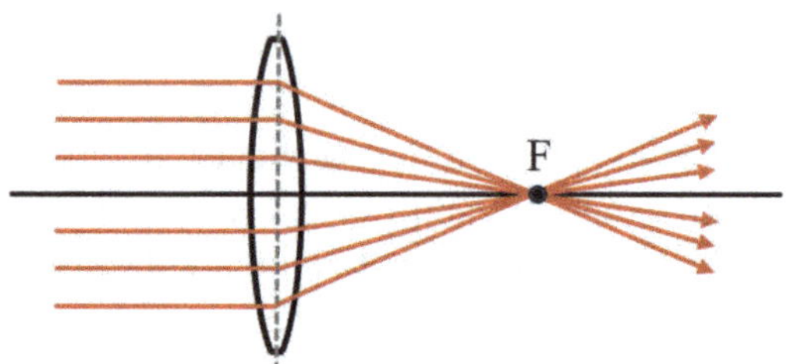

Converging Lens

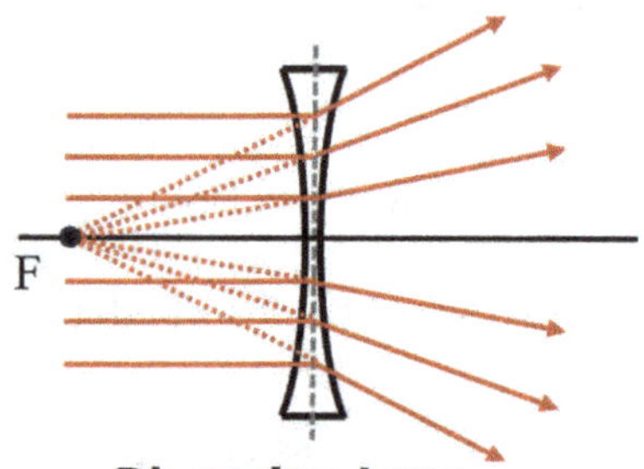

Diverging Lens

PROCEDURE

(1) Turn on the laser ray box and select 5 lasers.

(2) Use the various lenses to explore how a particular lens/mirror refracts or reflects the light.

(3) Complete Question 1.

LIGHT INTENSITY VS DISTANCE

BACKGROUND

As you move away from a light source, the light gets dimmer in accordance with the inverse square law, which states that the intensity of the light decreases with the square of the distance.

$$I \propto \frac{1}{d^2}$$

This results from geometry; therefore, the inverse square law applies to many phenomena, not just light. For example, point sources of electric field, light, sound or radiation obey the inverse square law.

PROCEDURE

(1) Clip the light source assembly to the side of the track. Position it so the LED is exposed and is facing along the length of the track. Align the light source assembly with the 10 cm mark on the track.

(2) Turn on the light source. It requires about **15 min** before it comes to thermal equilibrium, which we need for consistent measurements.

(3) Position the cradle that holds the light sensor on the base of the light sensor holder using the pins on the right side as you look towards the light source. The arrows on the cradle should point towards the light source. Place the light sensor in the cradle and snap it in place. This will align the light sensor with the light source in the center of the track.

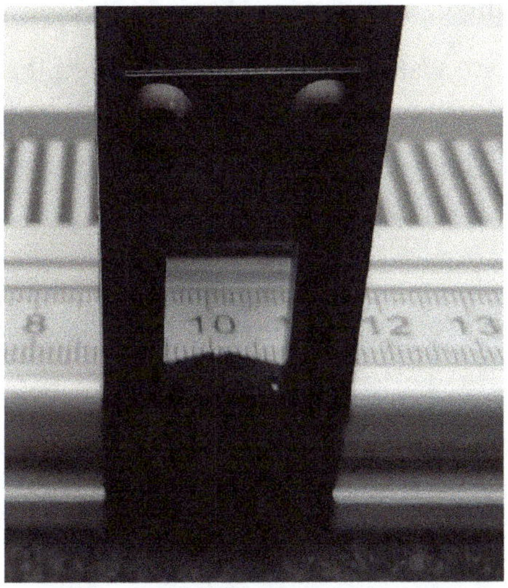

(4) Minimize any external lights (laptops, phones, lab lights, etc.).

(5) Start Graphical Analysis.

(6) Turn on the color & light sensor.

(7) Click Sensor Data Collection and select your sensor from the list of Discovered Wireless Devices.

(8) Click Sensor channels and select: Light, 615 nm, 525 nm, and 465 nm (white light, red, green and blue respectively). Click Done.

(9) Click View Options and select 1 graph. It should be Illumination vs Time.

(10) Click the label on the vertical axis (should say Illumination at this point) and select: Illumination, 615 nm, 525 nm, and 465 nm. Change colors to black, red, green, and blue respectively.

(11) Rotate the disk on the light source assembly until no light from the LED is visible. Zero the sensor. This minimizes the effect of background light.

(12) Rotate the disk on the light source so that light shines through once again.

(13) Position the light sensor so the notched arrow is at the 20 cm mark. This makes the distance between the light source and the sensor 10 cm.

(14) Click Collect to begin data collection. Stop after 5 seconds.

(15) Go to view statistics and record the mean in Table 1.

(16) Position the light sensor so the notched arrow is at the 30 cm mark. This makes the distance between the light source and the sensor 20 cm.

(17) Repeat the process until Table 1 is complete.

(18) Restart Graphical Analysis in Manual Entry mode.

(19) In the X column, enter the distance (10, 20, … 100).

(20) In the Y column, enter the Illumination for the white light.

(21) Add a manual column (click on the 3 dots in the Y column and select Add Manual Column)

(22) Rename it "Red" and enter the values for the red channel.

(23) Add columns for the green and blue channels and enter their values.

(24) Ensure the vertical axis has all four data series enabled so they are visible on the graph (click on the vertical axis label and select).

(25) Apply an inverse squared curve fit – it should do this for all four data sets at once. Click Apply.

(26) **DO NOT TURN OFF THE LIGHT SOURCE YET.**

(27) Report the RMSE in Table 1.

(28) Complete Question 2.

POLARIZATION OF LIGHT

BACKGROUND

Sunlight and artificial illumination produce electromagnetic waves whose electric field vectors vibrate randomly in all planes. If these vectors are restricted to a single plane, then the light is polarized.

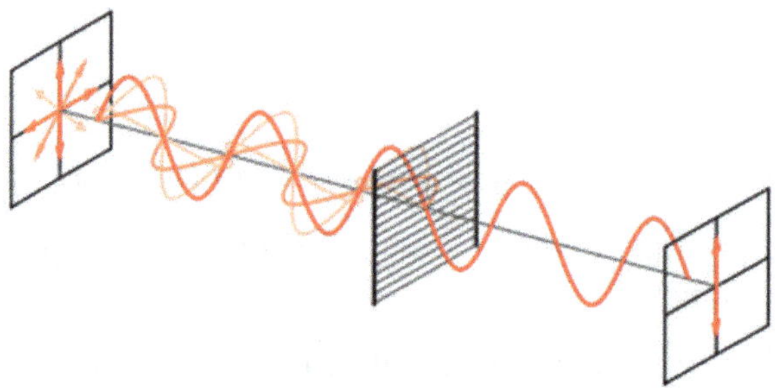

When two polarizing filters are used, we can expect two logical end points. If both filters are aligned (e.g., both are aligned vertically), the first filter will vertically polarize the light, and the second will transmit the vertically polarized light. On the other hand, if the polarizing filters are "crossed" (i.e., 90° to one another), then the second filter cannot pass the crossed polarized light. For intermediate positions (e.g., the second filter is rotated 30° with respect to the first) then various amounts of light are transmitted by the second filter. The intensity of the light is angle-dependent and is given by Malus's Law:

$$I = I_0 \cos^2(\theta)$$

where I is the intensity of the light passing through the polarizers, I_0 is the original intensity of the light, and θ is the angle between the polarizers' alignment.

PROCEDURE

(1) Using the previous setup, place two polarizers between the light source and the light sensor. Place the first polarizer close to the light source (about 10 cm away) with the dial of the polarizer on the side away from the light source.

(2) Set the polarizer to 0°. This polarizer will never change its setting.

(3) Place the second polarizer about 10 cm from the first one, also with the dial of the polarizer on the side away from the light source. Place this polarizer at 90°, which should stop all light. We will refer to this polarizer as the analyzer from now on.

(4) Place the color and light sensor about 5 cm from the analyzer.

(5) Start Graphical Analysis.

(6) Turn on the color & light Sensor (if needed).

(7) Click Sensor Data Collection and select your sensor from the list of Discovered Wireless Devices.

(8) Click Sensor channels and select Light. Click Done.

(9) Click View Options and select 1 graph. It should be Illumination vs Time.

(10) As before, zero the sensor to minimize the effect of background light.

(11) Rotate the analyzer to 0°.

(12) Click Collect to begin data collection. Stop after 5 sec.

(13) Go to view statistics and record the mean in Table 2.

(14) Rotate the analyzer to 30° and repeat the process.

(15) Repeat the process until Table 2 is complete.

(16) Restart Graphical Analysis in Manual Entry mode.

(17) In the X column, enter the angle (0, 30, … 360).

(18) In the Y column, enter the Illumination for the white light.

(19) Apply a cosine squared curve fit. Click Apply.

(20) Report the RMSE in Table 2.

(21) Complete Question 3.

LAB 14 – ASSIGNMENT

QUESTIONS – REFLECTION & REFRACTION

(1) Sketch the ray diagram for the following lenses.

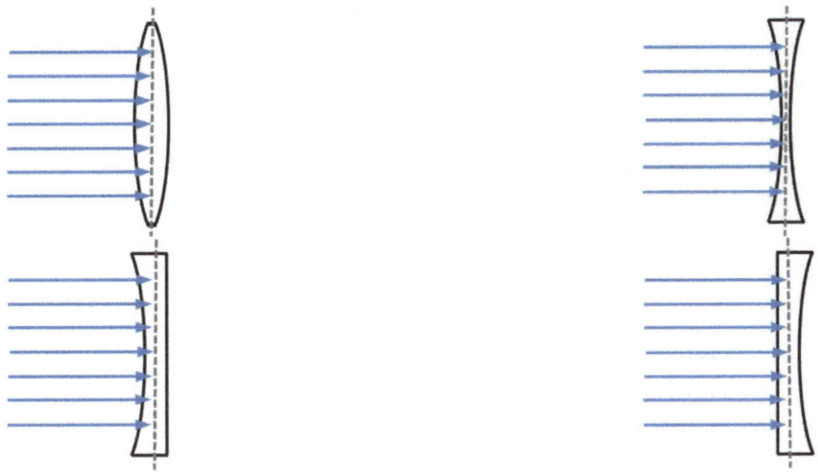

QUESTIONS – LIGHT INTENSITY VS DISTANCE

Table 1

Distance (cm)	Illumination (lux)	Red (lux)	Green (lux)	Blue (lux)
10				
20				
30				
40				
50				
60				
70				
80				
90				
100				
RMSE				

(2) If the intensity of light is 1200 lux at 0.2 m, what would be the intensity at a distance of 0.5 m?

QUESTIONS – POLARIZATION OF LIGHT

Table 1

Angle	Illumination (lux)
0	
30	
60	
90	
120	
150	
180	
210	
240	
270	
300	
330	
360	
RMSE	

(3) What angle is needed between the direction of polarized light and the axis of the polarizing filter to reduce its intensity by 85%?

LAB 15 | ELECTROSTATICS

GETTING STARTED

OVERVIEW

In this lab, you will observe electrical charge separation and transfer, and you will learn how to measure the electric charge of objects.

EQUIPMENT

- Lab Computer
- Charge Sensor
- Electrostatics Kit
- Carpet Square

ELECTROSTATIC CHARGE ON THE BODY

BACKGROUND

In cold or dry conditions (like we have here in the Spring Semester), your body can build up excess electrical charge simply from walking. Shoes often have insulating soles, which allow a static charge to accrue as they rub against the floor. This charge on your shoes induces a buildup of charge on your body, creating a high voltage. Human skin tends to be electropositive due to the presence of electrolytes (e.g., sodium, calcium, and magnesium), which readily lose electrons and contribute to a positive charge. Despite the high voltage, static shocks are usually not harmful because the skin has a relatively high electrical resistance. This limits the current during a discharge to very small levels, which is why a static shock may be briefly painful but not dangerous.

PROCEDURE

(1) Set up the Faraday pail, cage and charge sensor. Connect the black lead from the charge sensor to the grounding plane and connect the red lead from the charge sensor to the Faraday pail.

(2) Connect a grounding wire between the cage and the grounding plane.

CAUTION

- The sensor is capable of measuring very small amounts of charge. As the experiment proceeds, the sensor will likely develop a charge and it will be important to remove the excess charge. To do so, press and hold the Reset button on the sensor. If it is a cold or dry day, you will likely need to do this frequently.

(3) Start Graphical Analysis and click Sensor Data Collection.

(4) Turn on the charge sensor and select your sensor from the list of Discovered Wireless Devices.

(5) Configure the View for a graph, a data table, and a meter.

(6) Scuff your shoes on the carpet square, then insert a finger into the pail without touching either the pail or the cage. Observe the readings on the meter.

(7) Remove your finger. Again, observe the readings on the meter.

(8) Touch the metal grounding plane that holds the Faraday pail with your finger.

(9) Insert your finger into the pail without touching the pail. Observe the reading on the meter. Remove your finger. Again, note the readings on the meter.

(10) Ground the pail and zero the sensor (reset button).

(11) Attach the alligator clip on the grounding strap to the grounding plane and attach the grounding strap to your wrist. Repeat Steps 2 and 3.

(12) Complete Questions 1–2.

METHODS OF CHARGING

BACKGROUND

Charging means gaining or losing electrons thereby creating an imbalance. There are three ways for this to happen: charging by friction, charging by contact (conduction) and charging by induction. When you rub one material against another, they are charged by friction. One material loses electrons and becomes positively charged, while the other material gains electrons and becomes negatively charged.

Charging by conduction occurs when two objects with different amounts of electric charge come in contact and electrons move from one material to the other. If the charged object is negatively charged, electrons will move onto the neutral object. If the charged object is positively charged, electrons from the neutral object will move onto the positively charged object.

Charging by induction is a method to generate static charges in a material by bringing an electrically charged object near it. This causes electrical charges to be redistributed in the materials, resulting in one side having an excess of either positive (+) or negative (−) charges.

PROCEDURE

(1) Choose New Experiment from the File menu. Change the data collection duration to 60 seconds and the rate to 10 samples per second.

(2) As before, configure the View for a graph, a data table, and a meter.

(3) Attach the alligator clip on the grounding strap to the grounding plane and then attach the grounding strap to your wrist. Discharge any excess charge on the sensor using the reset button.

Charging by Friction

(4) Dab (*do not rub*) each charge separator on a damp cloth to remove any charge from the charge separators.

(5) Insert the charge separators into the Faraday pail one at a time to verify that they are electrically neutral (the meter will read 0 nC). **NOTE** – If it is a dry day (which is likely), it might be difficult to eliminate excess charges. Just get them as close to neutral as possible (this is a qualitative lab, not a quantitative one). The white charge separator seems to be more difficult to discharge.

(6) Rub the white and gray charge separators together, then hold them apart.

(7) **READ ALL OF THIS BEFORE PROCEEDING.** Click Collect to begin data collection. Insert the white charge separator into the middle of the Faraday pail without touching the pail. Wait 4 or 5 seconds, then remove it. Wait a couple of seconds, then insert the gray charge separator into the middle of the pail and wait 4 or 5 seconds. Remove it and wait a couple of seconds. Finally, position both charge separators in the pail; be careful to keep the charge separators from touching the pail or each other. Wait 4 or 5 seconds, then remove both separators and stop data collection.

(8) Sketch your graph (Question 3).

Charging by Contact

(9) Continue using the grounding strap. Discharge the charge sensor (reset button).

(10) To remove any charge from the charge separators, dab each charge separator on a damp cotton cloth (again, *no rubbing*). Insert the charge separators one at a time into the Faraday pail to verify that they are neutral.

(11) Rub the charge separators together, then set the gray one to the side.

(12) **READ ALL OF THIS BEFORE PROCEEDING.** Click Collect to begin data collection. Insert the white charge separators into the middle of the Faraday pail without touching the pail. Wait 4 or 5 seconds, then remove it. Wait a couple of seconds. Now lightly rub the charge separator across the top edge of the pail and remove it from the pail. Stop data collection.

(13) Sketch your graph (Question 6).

Charging by Induction

(14) Continue using the grounding strap. Discharge the charge sensor (reset button).

(15) Attach a second grounding wire to the grounding plane, leaving the other end disconnected (you'll use this later to ground the pail).

(16) To remove any charge from the charge separators, dab each charge separator on a damp cotton cloth (again, *no rubbing*). Insert the charge separators one at a time into the Faraday pail to verify that they are neutral.

(17) Rub the charge separators together, then set the gray one to the side.

(18) **READ ALL OF THIS BEFORE PROCEEDING.** Insert the charge separator into the middle of the pail (be careful to not touch the side of the pail). Wait 4 or 5 seconds, then remove it. Wait

another couple of seconds. Reinsert the charge separator into the pail, then ground the pail by briefly touching it with the second grounding wire. Wait 4 or 5 seconds, then remove the charge separator. Wait a couple of seconds and move the charge separator back into the pail. Wait 4 or 5 seconds, then remove the charge separator and stop data collection.

(19) Sketch your graph (Question 9)

(20) Complete Questions 3–10.

LAB 15 – ASSIGNMENT

QUESTIONS – ELECTROSTATIC CHARGE ON THE BODY

(1) What happened to the charge when you inserted and removed your finger from the Faraday pail?

(2) What effect did the ground strap have on charge?

QUESTIONS – METHODS OF CHARGING

(3) Sketch your graph for Charging by Friction.

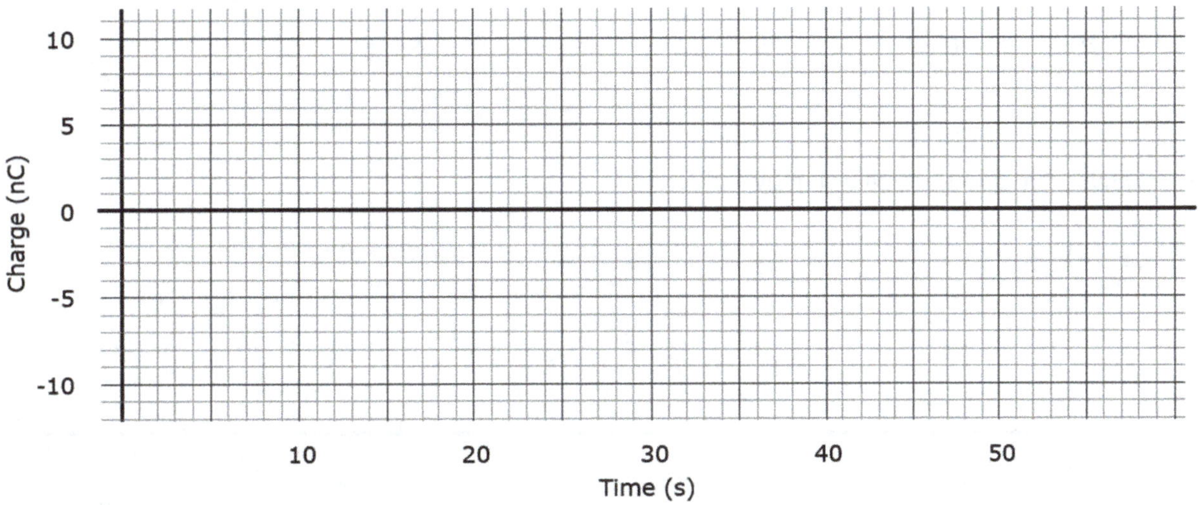

(4) Based on your results for the Charging by Friction portion, what charge (positive or negative) does the nylon charge separator (the white one) acquire and what charge does the gray separator acquire during friction? How do you know?

(5) When both charge separators were inserted in the Faraday pail, what happened to the magnitude of the charge? Why?

(6) Sketch your graph for Charging by Contact.

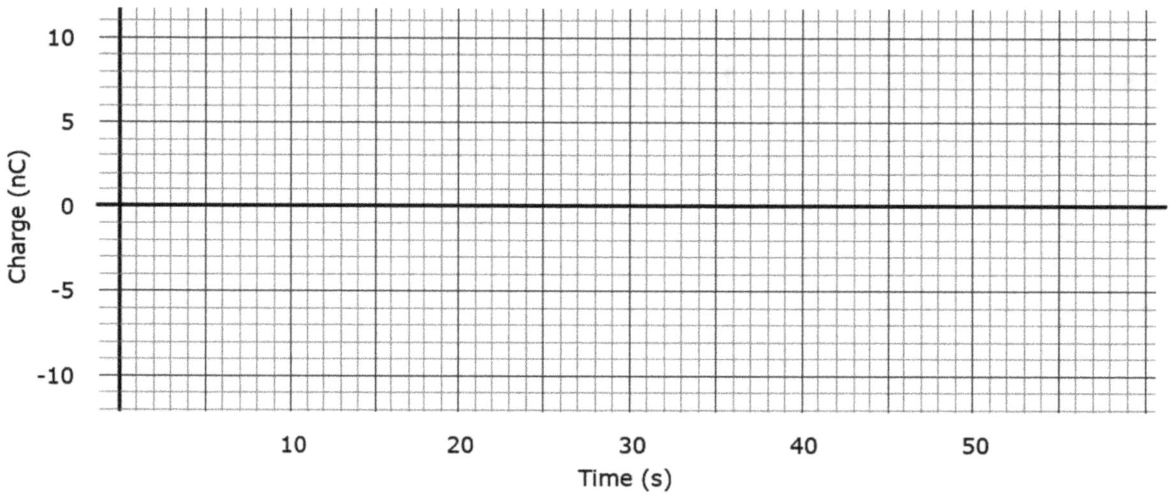

(7) Did the charge in the pail return to zero charge at the end of the experiment? Why or why not? Explain in terms of electrons and where they go in this process.

(8) How would the graph have changed if you used the gray charge separator instead?

(9) Sketch your graph for Charging by Induction.

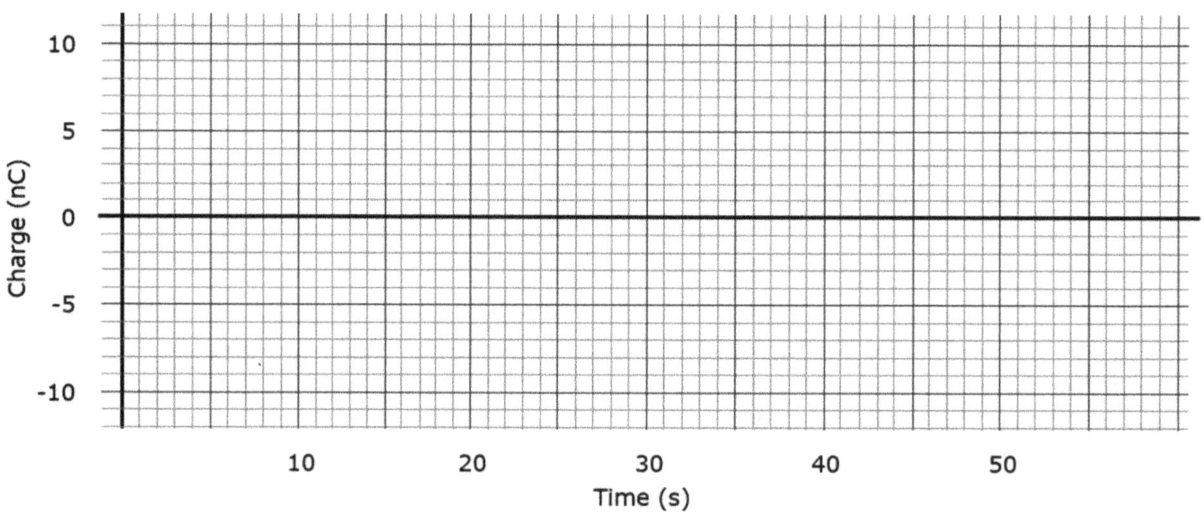

(10) What type of charge (positive or negative) did the white charge separator have? What type of charge did it induce on the pail? How do you know?

LAB 16 | ELECTRIC FIELDS

GETTING STARTED

OVERVIEW

In this lab, you will experimentally map electric field lines.

EQUIPMENT

- DC Power Supply
- DC Power Supply Cables
- Field Mapper Kit
- Digital Multimeter
- Multimeter Cables with Clips

ELECTRIC FIELD LINES & EQUIPOTENTIAL LINES

BACKGROUND

Every charged object creates an electric field, whose shape, direction, and strength determines how other charges will move when they enter it. Electric fields exist in three dimensions, which makes them tough to envision. To make things easier, physicists use a visual tool called electric field lines - imaginary lines that show the direction a *positive* test charge would move. By convention, electric field lines begin at positive charges and end at negative charges. The number of electric field lines drawn leaving a positive charge or approaching a negative charge is proportional to the magnitude of the charge and the field lines never cross because the electric field at any point has only one direction.

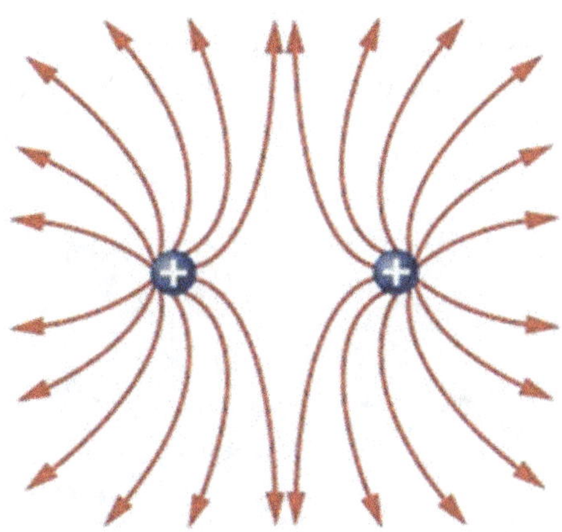

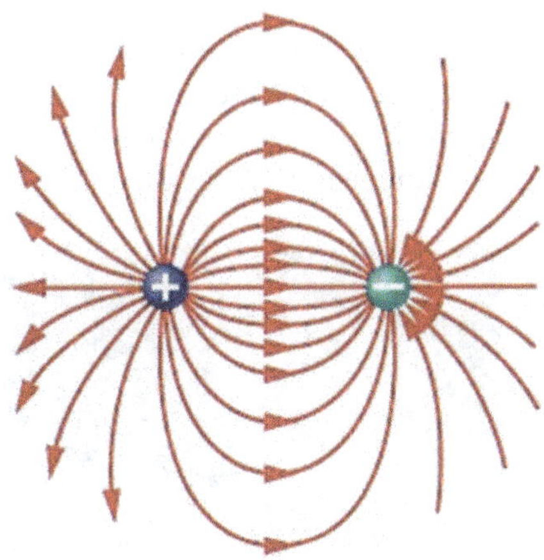

If all points in an electric field are at the same electric potential, they are equipotential points. If these points are connected by a line or a surface, we call it an equipotential line or surface respectively. If a point charge is moved from locations with different potentials (e.g., from V_A to V_B), the work done in moving the charge is given by:

$$W = q(V_A - V_B) = q\Delta V$$

The work done to move a point charge between two locations on an equipotential surface is zero since V_A and V_B are equal. Like electric field lines, equipotential lines will never intersect, but they are always perpendicular to the electric field lines. The spacing between equipotential surfaces indicates regions of strong or weak fields.

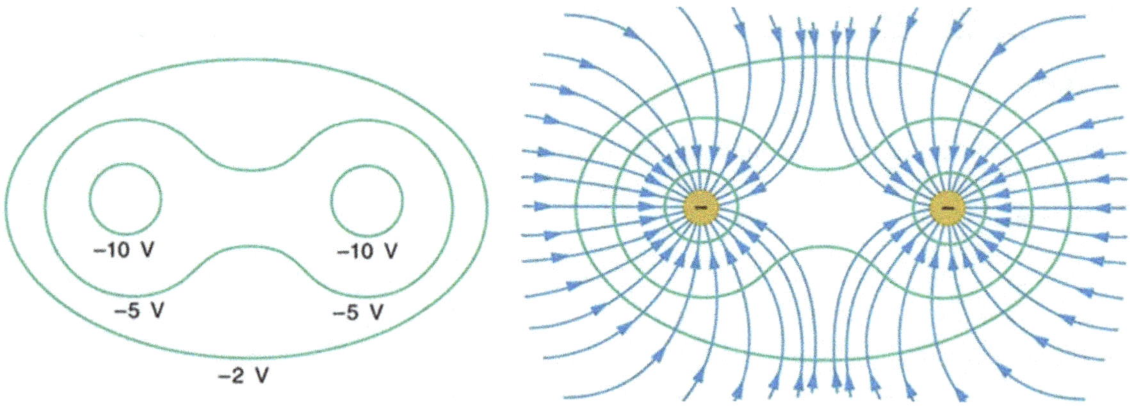

PROCEDURE

(1) Pin a sheet of conductive paper to the cork board using metal pushpins. There should be an electrode drawn on the paper with silver conductive ink.

(2) Press a metal pushpin into the center of each electrode, making sure that the pushpins are pressed through the paper and into the corkboard.

(3) With the power supply off, connect the electrode pushpins to the terminals on the power supply using two patch cords and alligator clips as shown: positive terminal to one electrode (this is the positive electrode), negative terminal to the other (this is the negative electrode).

(4) Connect the two remaining patch cords to the DC voltage ports on the digital multimeter (DMM) and then adjust the DMM to measure DC voltage (V_{DC} setting). Change the Range to Volts (the default is millivolts) by pressing the range button twice. If you get a reading of OL (out of limits) later, then you need to press the range button once more.

(5) Connect the DMM "ground" or "COM" patch cord to the pushpin in the negative electrode (this is now the reference electrode). The other patch cord will be used as a probe.

(6) On the graph paper, draw in the electrodes.

(7) Turn on the power supply and adjust the voltage to 10 V. Adjust the current so that it is not zero.

(8) Touch the tip of the probe to any black space surrounding either electrode on the conductive paper and observe the voltage measurement on the DMM. It should measure different voltages at different points on the paper. If not, make sure all the clips and pushpins are making good connections and retest. ***NOTE***: Touch the tip of the probe only to the solid black areas of the conductive paper; don't touch the paper's grid marks.

(9) Use the tip of the probe to touch different locations surrounding the electrodes until you find a location on the paper where the DMM reads 1.0 V (or something very close). Use the grid markings to locate this position on your graph paper and make a small mark.

(10) Continue moving the probe to different points on the paper, identifying several locations surrounding the electrodes where the voltage is at 1.0 V. Mark each new point on your graph paper until there are enough marks to draw a smooth line that connects them. Label the line "1.0 V". ***NOTE***: Be sure to probe all areas surrounding the electrodes. Isolines of electric potential may connect in a closed path on the paper or extend off the edge of the paper and then re-enter the paper in unexpected locations.

(11) Repeat the previous data collection steps, identifying and drawing the isolines of electric potential for each of these voltage values: 1, 2, 3, 4, 5, 6, 7, 8, and 9 V. Label each line with its voltage.

(12) Turn the power supply off when you are finished.

(13) Complete Questions 1–5.

LAB 16 – ASSIGNMENT

QUESTIONS – ELECTRIC FIELD LINES & EQUIPOTENTIAL LINES

Use the graph paper (next page) to map the equipotential lines as described in the Procedure.

(1) Once the equipotential lines have been drawn, draw in the electric field lines. Continue making electric field lines until there are at least 10 of them.

(2) A charged particle (q =1.4 nC) moves 0.24 m along an equipotential line of 9 V. Calculate the work done by the electric field during this motion.

(3) A charged particle of charge 1.2 nC accelerates in a uniform electric field of 100 V/m. The particle started from rest on an equipotential plane of 40 V. After t = 0.002 seconds, the particle is on an equipotential plane of V = 10 volts. Determine the distance traveled by the particle.

(4) A charged particle ($m = 5\cdot 10^{-20}$ kg) of charge 1.2 nC is accelerated from rest through a potential difference of 100 V. What is the speed of the particle at the end of the acceleration?

(5) If we placed an electron at coordinates (6, 6) in the map from the lab, which direction would it go? Use one of the 8 cardinal directions (N, S, E, W, NE, NW, SE, SW).

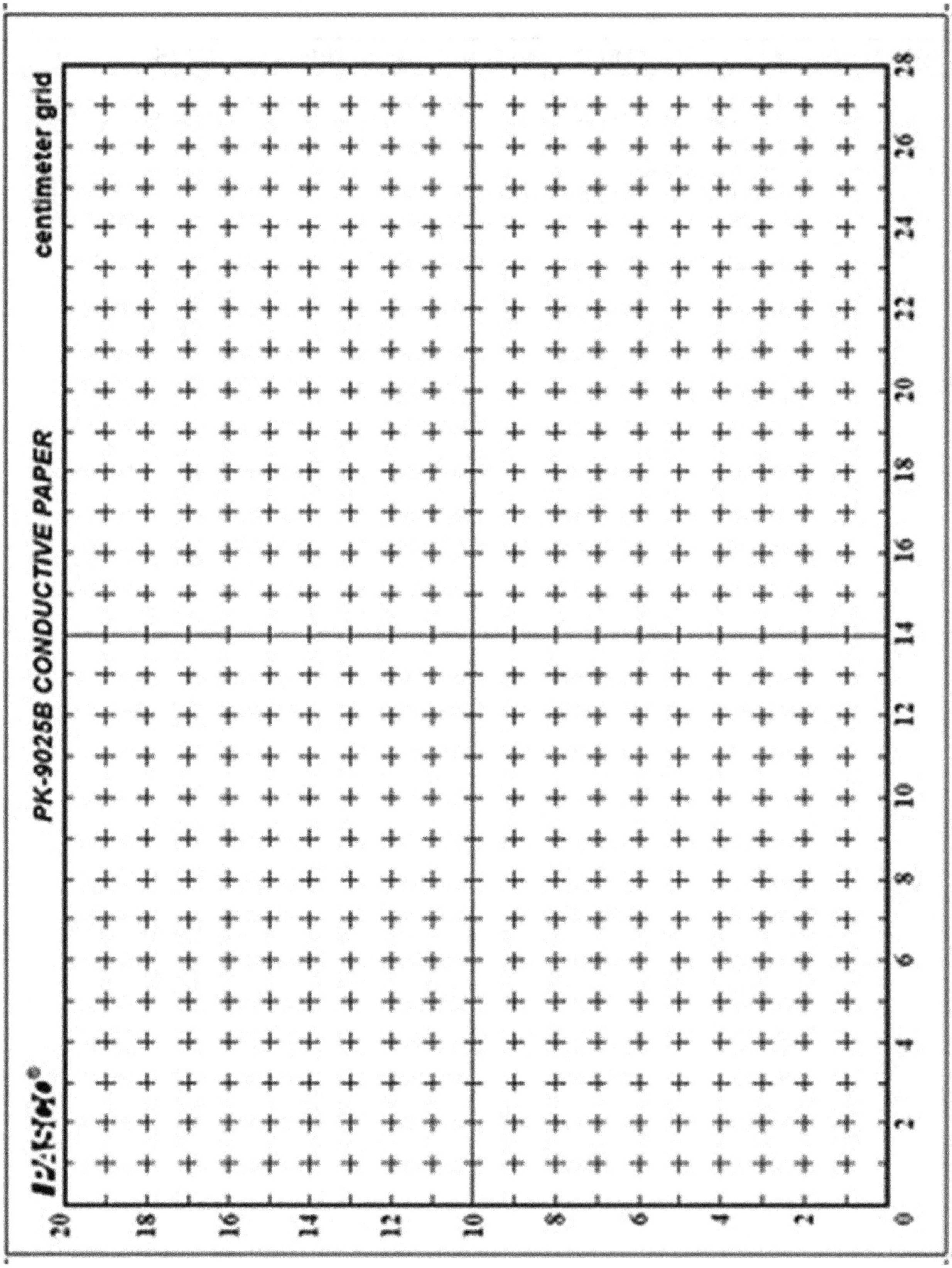
PASCO®
PK-9025B CONDUCTIVE PAPER
centimeter grid

LAB 17 | CAPACITANCE

GETTING STARTED

OVERVIEW

In this lab, you will explore the concept of capacitance and its relationship with voltage.

EQUIPMENT

- Lab Computer
- Modular Circuit Kit
- Voltage Sensor

VOLTAGE IN CAPACITOR CIRCUITS

BACKGROUND

Capacitors are electronic devices that store electrical energy in an electric field between two conducting plates that are separated by a dielectric material. Capacitance is a measure of a device's ability to store electrical charge per unit of voltage. The SI unit for capacitance is the Farad.

$$C = \frac{Q}{V}$$

When in series, capacitors are connected end-to-end so that the electrical charge is the same for each. The total voltage supplied is divided among the capacitors. If the capacitors have the same value, each will have an equal share of the voltage; otherwise, the voltage is proportional. For capacitors connected in series, the equivalent capacitance (C_{eq}) is:

$$C = \frac{Q}{V} \quad \Rightarrow \quad V = V_1 + V_2 = \frac{Q}{C_{eq}} = \frac{Q}{C_1} + \frac{Q}{C_2} \quad \Rightarrow \quad \frac{1}{C_{eq}} = \frac{1}{C_1} + \frac{1}{C_2} \quad \Rightarrow \quad C_{eq} = \left(\frac{1}{C_1} + \frac{1}{C_2} \right)^{-1}$$

or more generally,

$$C_{eq} = \left(\frac{1}{C_1} + \frac{1}{C_2} + \ldots + \frac{1}{C_n} \right)^{-1} = \sum_{i=1}^{n} \left(\frac{1}{C_i} \right)^{-1}$$

When arranged in parallel, capacitors are connected at both ends and share the same voltage since the current flows through each parallel branch independently. Each capacitor charges separately. Capacitors connected in parallel have an equivalent capacitance of:

$$Q_{total} = Q_1 + Q_2 = C_1 V + C_2 V = (C_1 + C_2) V \quad \Rightarrow \quad C_{eq} = C_1 + C_2$$

or more generally,

$$C_{eq} = C_1 + C_2 + \ldots + C_n = \sum_{i=1}^{n} C_i$$

PROCEDURE

(1) Build the following circuit with the modular circuit components. Don't forget to connect the blocks electrically with the metal tabs.

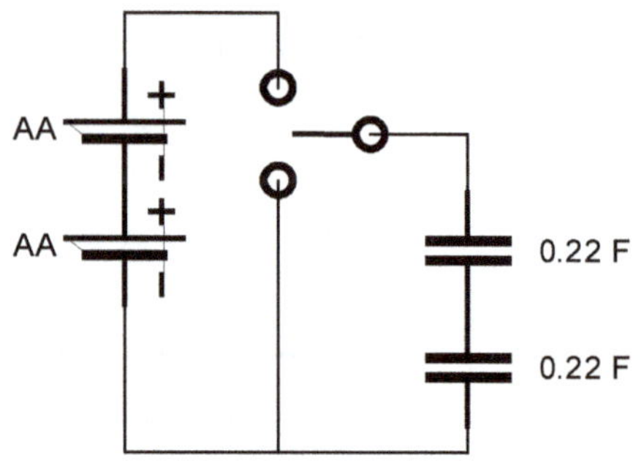

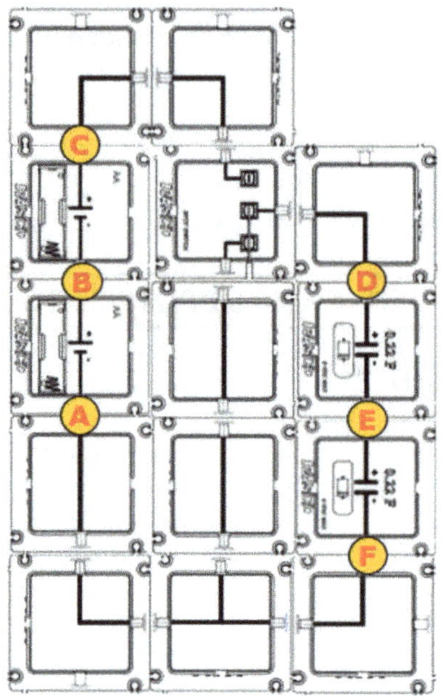

(2) Turn on the voltage sensor.

(3) Start the Vernier Graphical Analysis application.

(4) Click Sensor Data Collection and select your sensor from the list of Discovered Wireless Devices.

(5) Configure for 1 graph (Voltage vs. Time).

(6) Zero the voltage sensor.

(7) Place the switch so that it does not connect at all (straight up).

CAUTION

- Always connect the positive lead (red) to the positive end of the components.

(8) Clip the leads of the voltage sensor across the lower battery (positions A and B in the diagram); so, red goes to position B (positive terminal) and black to position A (negative terminal).

(9) Click Collect to begin data collection. Collection should stop after 5 seconds.

(10) On the graph, highlight an area where the voltage is steady and find the mean (View Statistics). Record this value in Table 1.

(11) Disconnect and zero the voltage sensor.

Capacitors in Series

(12) Position the switch to place voltage across the capacitors (switch to the top).

(13) Wait 2 minutes for the voltage in the circuit to stabilize.

(14) Repeat steps 8–10 to measure the voltage across C1 (DE). If you get a negative voltage – reverse the leads. Record the value in Table 2.

(15) Measure the voltage across C2 (EF) and record the value in Table 2.

(16) Measure the voltage across both capacitors (DF) and record the value in Table 2.

Capacitors in Parallel

(17) Build the following circuit with the modular circuit components.

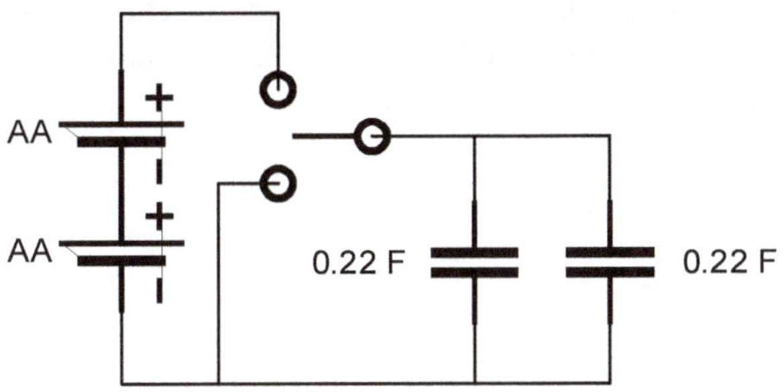

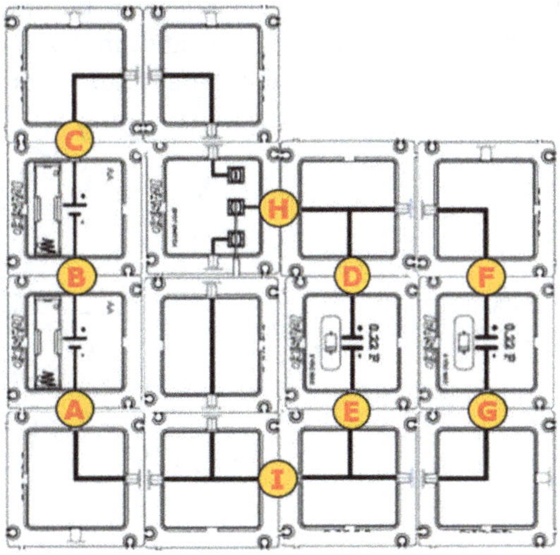

(18) Position the switch to place voltage across the capacitors (switch to the top).

(19) Wait 2 minutes for the voltage in the circuit to stabilize.

(20) Repeat steps 8–10 to measure the voltage across C1 (DE). Record the value in Table 3.

(21) Measure the voltage across C2 (FG) and record the value in Table 3.

(22) Measure the voltage across both capacitors (HI) and record the value in Table 3.

(23) Disconnect the voltage sensor.

(24) Complete Questions 1–2.

VOLTAGE VS PLATE SEPARATION IN CAPACITORS

BACKGROUND

Most capacitors consist of a pair of parallel plates, which are thin sheets of metal separated by a dielectric (insulating) material. For this reason, the symbol for a capacitor is two parallel lines (representing the capacitor plates). Usually, the plates are rolled up in a spiral to increase the overall area while keeping the device compact. Capacitance is given by:

$$C = \frac{\kappa \varepsilon_0 A}{d}$$

where κ is the dielectric constant of the spacer material, ε_0 is the permittivity of free space ($8.85 \cdot 10^{-12}$ F/m), A is the area of one of the metal plates ($2.46 \cdot 10^{-2}$ m^2), and d is their separation distance.

An experiment was done using a variable capacitor, which allows the distance between the plates to vary. The voltage across a capacitor was measured at various plate separation distances. The variable capacitor was connected to a 30 V source and the voltages on each plate measured with an electrometer (essentially a voltage sensor). Since

$$C = \frac{Q}{V} \Rightarrow \frac{Q}{V} = \frac{\kappa \varepsilon_0 A}{d} \Rightarrow V = \frac{Qd}{\kappa \varepsilon_0 A}$$

we see that as the separation distance increases, voltage also increases in a linear relationship. However, the system used in the experiment was a bit more complex. Other components in the experimental setup have some capacitance (C_{sys}), because they are part of the circuit, unlike when we used the voltage sensor earlier – it was not part of the circuit. So, we must account for the extra capacitance:

$$C = \frac{\kappa \varepsilon_0 A}{d} + C_{sys} \Rightarrow V = \frac{Q}{\left(\frac{\kappa \varepsilon_0 A}{d}\right) + C_{sys}}$$

For the experiment:

$A = 2.46 \cdot 10^{-2}$ m^2	Given (measured)
$\kappa = 1.0$	From a reference table
$\varepsilon_0 = 8.85 \cdot 10^{-12}$ F/m	Constant
$C_{sys} = 2.7 \cdot 10^{-11}$ F	Experimentally determined
$Q = 2.2 \cdot 10^{-9}$ C	Experimentally determined

Here is the other experimental data:

Separation (m)	Voltage (V)	Separation (m)	Voltage (V)	Separation (m)	Voltage (V)
0.003	30.42	0.020	67.73	0.060	77.91
0.005	43.45	0.030	72.40	0.070	79.49
0.010	56.64	0.040	75.04	0.080	81.11
0.015	63.46	0.050	76.70		

So, let's "simplify" the equations above.

$$\kappa \varepsilon_0 A = (1)\left(8.85 \cdot 10^{-12}\ \text{F/m}\right)\left(2.46 \cdot 10^{-2}\ \text{m}^2\right) = 2.18 \cdot 10^{-13}\ \text{F} \cdot \text{m}$$

$$V = \frac{Q}{\left(\frac{\kappa \varepsilon_0 A}{d}\right) + C_{sys}} = \frac{2.2 \cdot 10^{-9}\ \text{C}}{\left(\frac{2.18 \cdot 10^{-13}\ \text{F} \cdot \text{m}}{d}\right) + 2.7 \cdot 10^{-11}\ \text{F}}$$

PROCEDURE

(25) Use the simplified equation to predict the voltage for each distance. Place the values in Table 4.

(26) Start Graphical Analysis.

(27) Select Manual Data Entry and enter the data from Table 4 into the data table. Use the separation distance as the x column and the voltage as the y column.

(28) Add a new column (click on the 3 dots next to the Y and select Add Manual Column). Name the new column Z and enter the calculated data from Table 4.

(29) Configure the view for 2 graphs: experimental voltage vs. distance (Y vs X) and predicted voltage vs. distance (Z vs X).

(30) Fit a natural Exponent curve to both plots. Report the equations and RMSE in Q3 & Q4 respectively.

(31) Complete Questions 3 – 5.

VOLTAGE VS DIELECTRIC CONSTANT

BACKGROUND

Dielectric materials are nonconductive, so any dielectric material placed between the plates of a capacitor will increase its capacitance. Dielectric materials are used to keep the plates in a capacitor insulated from each other (preventing them from touching and consequently discharging). The dielectric constant for air is $\kappa = 1.00$. For all other materials, $\kappa > 1$.

Another experiment was done to determine the dielectric constant for a specific polystyrene plastic. This experiment had the same parameters, but a different setup; one where the system capacitance was not a factor. So,

$$C = \frac{\kappa \varepsilon_0 A}{d} \quad \Rightarrow \quad V = \frac{Q}{\left(\frac{\kappa \varepsilon_0 A}{d}\right)} = \frac{Qd}{\kappa \varepsilon_0 A}$$

In this part of the experiment, the separation between the plates was fixed at 0.08 m and several voltage measurements were taken.

For this part of the experiment:

	$A = 2.46 \cdot 10^{-2}$ m^2	Given (measured)
	$d = 0.08$ m	Fixed
	$\varepsilon_0 = 8.85 \cdot 10^{-12}$ F/m	Constant
	$Q = 2.2 \cdot 10^{-9}$ C	Experimentally determined

Trial	1	2	3	4	Avg
Voltage (V)	315.1	316.9	317.2	314.5	315.9

PROCEDURE

(32) Complete Question 6.

LAB 17 – ASSIGNMENT

QUESTIONS – VOLTAGE IN CAPACITOR CIRCUITS

Table 1

	Battery 1 (AB)	Battery 2 (BC)	Calculated (AC)	Measured (AC)	|Difference|*
Voltage (V)					

*Absolute value between measured and calculated voltage.

Table 2

SERIES	C1 (DE)	C2 (EF)	Calculated (DF)	Measured (DF)	|Difference|*
Voltage (V)					

*Absolute value between measured and calculated voltage.

Table 3

PARALLEL	C1 (DE)	C2 (FG)	Calculated (HI)**	Measured (HI)	|Difference|*
Voltage (V)					

*Absolute value between measured and calculated voltage. ** Use the mean of C1 and C2.

(1) In which configuration, series or parallel, does an individual capacitor have the larger value?

(2) Suppose we have a circuit with an older 9 V battery, which is putting out 8.1 V. If we connect three capacitors in series, what will the voltage be across the middle capacitor?

QUESTIONS – VOLTAGE VS PLATE SEPARATION IN CAPACITORS

Table 4

Separation (m)	Voltage (V)	Predicted Voltage (V)	Separation (m)	Voltage (V)	Predicted Voltage (V)
0.003	30.42		0.040	75.04	
0.005	43.45		0.050	76.70	
0.010	56.64		0.060	77.91	
0.015	63.46		0.070	79.49	
0.020	67.73		0.080	81.11	
0.030	72.40				

(3) What was the equation for the natural exponent fit to the data (experimental voltage vs distance)? Also give the RMSE.

(4) What was the equation for the natural exponent fit to the data (predicted voltage vs distance)? Also give the RMSE.

(5) The existence of capacitance in the system (C_{sys}) really complicated things. What would a graph of the relationship between V and d look like if there were no system capacitance?

QUESTIONS – VOLTAGE VS DIELECTRIC CONSTANT

(6) Use the average voltage to compute the dielectric constant.

LAB 18 | RESISTANCE & OHM'S LAW

GETTING STARTED

OVERVIEW

In this lab, you'll explore the relationships between voltage, current, and resistance by experimentally verifying Ohm's Law.

EQUIPMENT

- Lab Computer
- DC Power Supply
- DC Power Supply Cables
- Voltage Sensor
- Current Sensor
- Vernier Circuit Board 2

RESISTANCE

BACKGROUND

Resistors are electronic devices that restrict the flow of electrical current. Resistance is a measure of this restriction and is defined as voltage divided by the current. The SI unit for resistance is the Ohm (Ω).

$$R = \frac{V}{I}$$

When in series, resistors are connected end-to-end and the same current will pass through each resistor along the only path available. However, the voltage is not the same across each resistor because each one removes some electrical energy, producing a voltage drop. The equivalent resistance is found by adding all the individual resistances together. Consequently, the equivalent resistance is always greater than any individual resistor in the series. For resistors connected in series, the equivalent resistance (R_{eq}) is:

$$V = V_1 + V_2 \quad \Rightarrow \quad V = IR_1 + IR_2 = I(R_1 + R_2) \quad \Rightarrow \quad V = IR_{eq} \quad \Rightarrow \quad R_{eq} = R_1 + R_2$$

or more generally,

$$R_{eq} = R_1 + R_2 + \ldots + R_n = \sum_{i=1}^{n} R_i$$

When arranged in parallel, resistors are connected at both ends and share the same voltage because each branch is attached directly across the same two points of the voltage source. However, the current is not the same in each branch because it depends on the resistance of each resistor. Recall that current flow is inversely proportional to resistance. For resistors connected in parallel, the equivalent resistance (R_{eq}) is:

$$I = I_1 + I_2 \quad \Rightarrow \quad \frac{V}{R_{eq}} = \frac{V}{R_1} + \frac{V}{R_2} \quad \Rightarrow \quad \frac{V}{R_{eq}} = V\left(\frac{1}{R_1} + \frac{1}{R_2}\right) \quad \Rightarrow \quad \frac{1}{R_{eq}} = \frac{1}{R_1} + \frac{1}{R_2} \quad \Rightarrow \quad R_{eq} = \left(\frac{1}{R_1} + \frac{1}{R_2}\right)^{-1}$$

or more generally,

$$R_{eq} = \left(\frac{1}{R_1} + \frac{1}{R_2} + \ldots + \frac{1}{R_n}\right)^{-1} = \sum_{i=1}^{n}\left(\frac{1}{R_i}\right)^{-1}$$

Resistors are often combined in series and in parallel within the same circuit (combination or complex circuit). To find the equivalent resistance in a complex circuit, you'll need to iteratively reduce the circuit. Break it into sections that are strictly series or parallel, find the equivalent resistance for that section, then use it for the next calculation.

PROCEDURE

(1) Complete Questions 1–2.

OHM'S LAW

Ohm's law describes the relationship between electrical current (I), voltage (V) and resistance (R):

$$I = \frac{V}{R} \quad \Rightarrow \quad V = IR \quad \Rightarrow \quad R = \frac{V}{I}$$

The law states that, in an ideal conductor, the electrical current is directly proportional to the potential difference (voltage) and inversely proportional to the resistance. Devices that obey Ohm's Law are called "ohmic" or "linear", because the voltage across the material varies linearly with current. Materials that do not follow Ohm's Law are "non-ohmic". In these devices, resistance typically increases as the applied voltage increases.

Current and voltage can be difficult to understand because they cannot be observed directly. To help clarify the ideas, we sometimes use a (flawed) analogy between electrical circuits and water flowing in pipes.

Electrical Quantity	Description	Unit	Water Analogy
Voltage	a measure of the energy difference per unit charge between two points	Volt (V)	Water pressure
Current	a measure of the flow of charge	Ampere (A)	Amount of water flowing
Resistance	a measure of how difficult it is for current to flow	Ohm (Ω)	A measure of how difficult it is for water to flow through a pipe

PROCEDURE

(1) Set up the experimental circuit and sensors as shown. Leave the power supply OFF. Connect across the 10 Ω resistor (pins 2 & 3). Note that there is a connector between pins 1 & 2 (green wire).

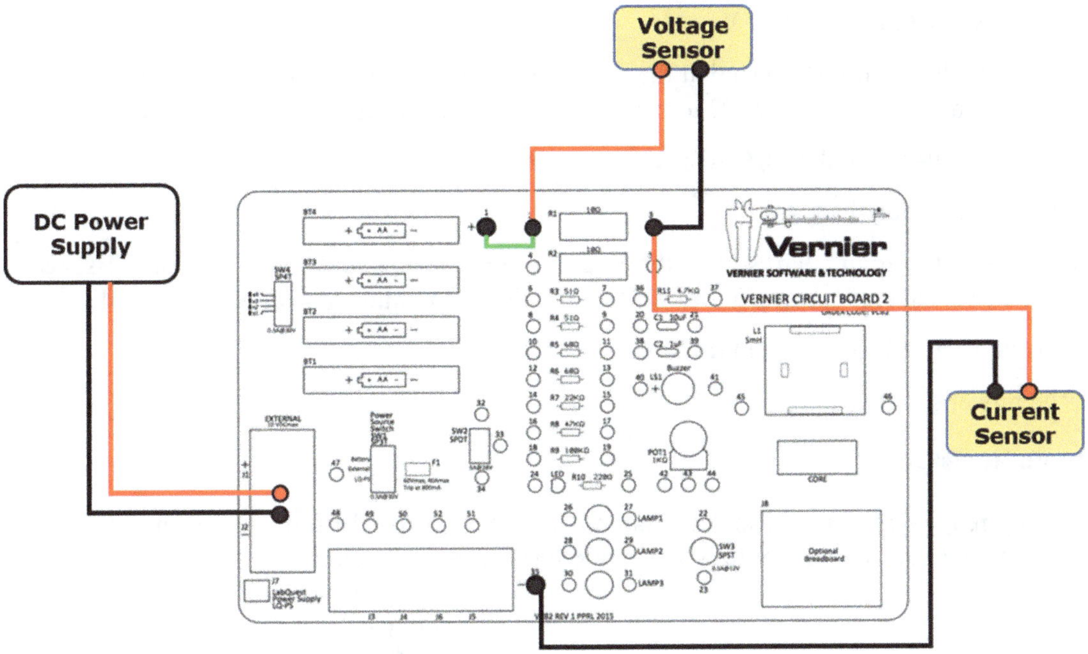

(2) On the circuit board, the power switch source should be set to external (it is likely already in this position).

(3) Turn on the current and voltage sensors. Ensure that the current sensor's switch is set to ± 1 A max.

(4) Start Graphical Analysis.

(5) Click Sensor Data Collection and select your sensor from the list of Discovered Wireless Devices.

(6) In the Data Collection Settings, change the mode to Event Based and then check Events with Entry.

(7) On the power supply, ensure that the voltage control is turned full CCW (off) and that it reads zero, then turn the current knob about 1/8 of a turn (it will continue to read zero since there is no voltage).

(8) Slowly increase the voltage to 4 V while monitoring the current readout. If you get to 4 V without a red warning light (current limited), then you are OK. If you get the warning, turn the current knob CW a small amount until the light goes away.

(9) Bring the voltage back to zero, leave the current alone.

Ohmic Resistance

(10) Click Collect to begin data collection. Click Keep – Keep Entry. This first reading should be 0 for both current and voltage.

(11) Bring the voltage on the DC power supply up to 0.5 V (note that the voltage sensor will likely read a little less than the voltage indicated on the power supply readout). Click Keep.

(12) Increase the voltage to 1.0 V and click Keep. Continue collecting data in 0.5 V increments until you reach 4 V. There should be a total of 9 readings. Click Stop.

(13) Record your data in Table 1 (10 Ω).

(14) On the graph, do a linear curve fit and report the data in Table 2.

(15) Set the power supply back to 0 V.

(16) Reconfigure the circuit board so that the 51 Ω resistor is connected rather than the 10 Ω resistor (i.e., move the two connections at pin 2 to pin 6, and the two connections at pin 3 to pin 7.

(17) Repeat steps 10–14 for the 51 Ω resistor.

(18) Set the power supply back to 0 V.

(19) Reconfigure the circuit board so that the 68 Ω resistor is connected rather than the 51 Ω resistor (i.e., move the two connections at pin 3 to pin 10, and the two connections at pin 7 to pin 11.

(20) Repeat steps 10–14 for the 68 Ω resistor.

(21) Set the power supply back to 0 V.

Non-Ohmic Resistance

(22) Reconfigure the circuit board so that a lamp is connected (pins 26 and 27). Ensure there is a light bulb in the holder. The light bulb will provide resistance in the circuit.

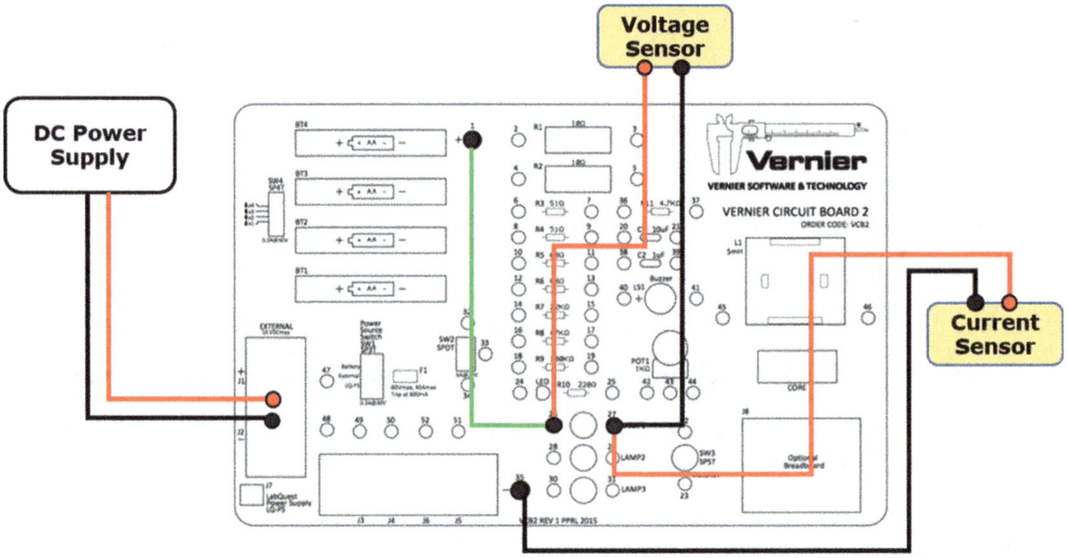

(23) Repeat steps 10–14 for the lamp, except do a power fit rather than a linear fit.

(24) Complete Questions 3–9.

LAB 18 – ASSIGNMENT

QUESTIONS – RESISTANCE

(1) Find the equivalent resistance in the circuit if: $R_1 = 3\,\Omega$, $R_2 = 2\,\Omega$, $R_3 = 5\,\Omega$, $R_4 = 6\,\Omega$, $R_5 = 7\,\Omega$.

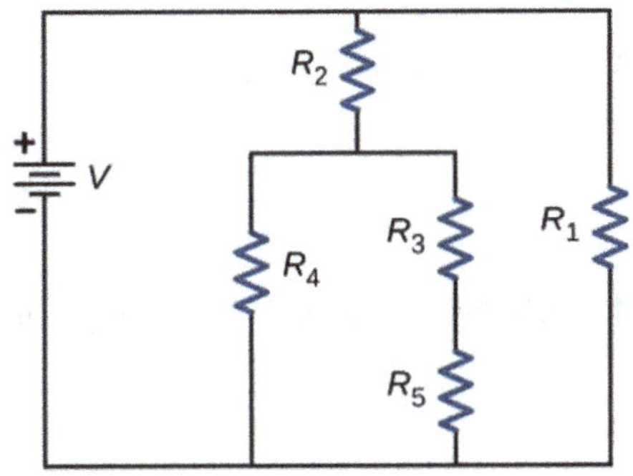

(2) Using the circuit in Q1, what value does R5 need to be to give the circuit an equivalent resistance of 4.5 Ω?

QUESTIONS – OHM'S LAW

Table 1

	10 Ω		51 Ω		68 Ω		Light Bulb	
Trial	**Voltage**	**Current**	**Voltage**	**Current**	**Voltage**	**Current**	**Voltage**	**Current**
1								
2								
3								
4								
5								
6								
7								
8								
9								

Table 2

	Slope (V/A)	Intercept (V)	Correlation
10 Ω Resistor			
51 Ω Resistor			
68 Ω Resistor			

(3) Ideally, what should the intercept be for any of the circuits? Why?

(4) Ideally, what should the slope be for the 10 Ω, the 51 Ω, and the 68 Ω resistor (don't forget units)?

(5) Do the resistors follow Ohm's Law? How do you know?

(6) Does the light bulb follow Ohm's Law? How do you know?

(7) What was the equation and correlation for your best fit to the light bulb data?

(8) Using (5), estimate the voltage if 0.2 A were flowing in the circuit.

(9) Using (5), estimate the current if we were to increase the voltage to that of a 9 V battery.

LAB 19 | DC CIRCUITS

GETTING STARTED

OVERVIEW

In this lab, you will continue to examine resistance for resistors connected in series, in parallel, and series-parallel circuits. In addition, you will explore charging and discharging a capacitor in an RC circuit.

EQUIPMENT

- Lab Computer
- Voltage Sensor
- Current Sensor
- Modular Circuits Kit

RESISTORS IN SERIES AND PARALLEL

BACKGROUND

The equivalent resistance of a combination of resistors depends on both their individual values and how they are connected. As we saw in the previous lab, the equivalent resistance for series and parallel circuits are respectively:

$$R_s = R_1 + R_2 + R_3 + \cdots + R_n = \sum_{i=1}^{n} R_i \qquad \text{and} \qquad R_p = \left(\frac{1}{R_1} + \frac{1}{R_2} + \frac{1}{R_3} \cdots + \frac{1}{R_n} \right)^{-1} = \left(\sum_{i=1}^{n} \frac{1}{R_i} \right)^{-1}$$

$$R_s = R_1 + R_2 + R_3 + \cdots + R_n = \sum_{i=1}^{n} R_i \qquad \text{and} \qquad R_p = \left(\frac{1}{R_1} + \frac{1}{R_2} + \frac{1}{R_3} \cdots + \frac{1}{R_n} \right)^{-1} = \left(\sum_{i=1}^{n} \frac{1}{R_i} \right)^{-1}$$

More complex connections of resistors are often just combinations of series and parallel connections. For example, this circuit has resistors that are in series and in parallel.

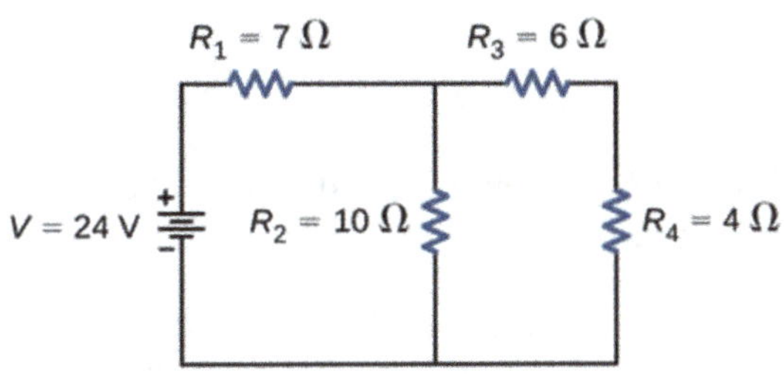

Such combinations are common and can be reduced to a single equivalent resistance as various portions can be identified as either series or parallel connections, reduced to their equivalent resistances, and then further reduced until a only single equivalent resistance is left.

PROCEDURE

(1) Turn on the voltage and current sensors.

(2) Start the Vernier Graphical Analysis application.

(3) Click Sensor Data Collection and select your sensor from the list of Discovered Wireless Devices.

(4) Under View Options, click Meters. This will provide live readouts of the sensors rather than relying on collecting data over time.

CAUTION

- Keep the switch in the off position until you need flow. This will help preserve the battery life.

Series Circuit

(5) Build a circuit with two 33 Ω resistors in series using modular components. Don't forget to connect them with the tabs as well.

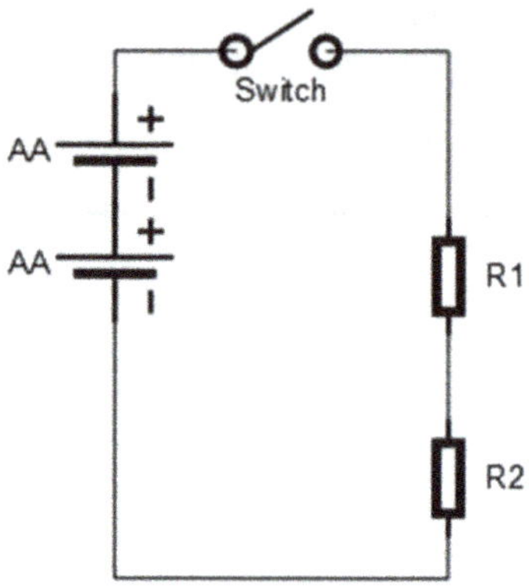

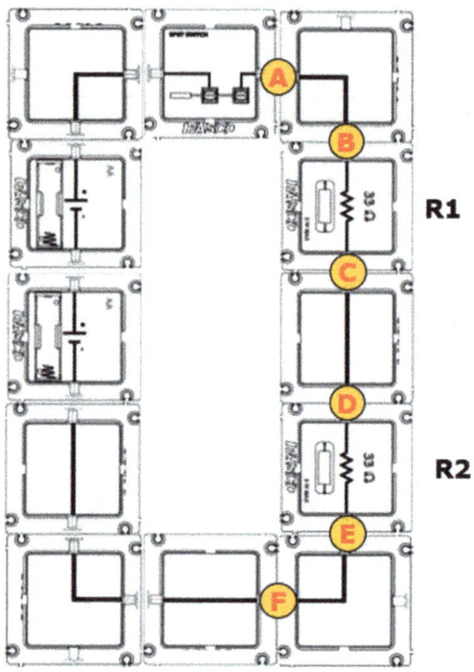

(6) Zero both sensors.

(7) Measure the voltage across R1 by connecting the red lead of the voltage sensor to B and the black lead to C. Record the data in Table 1.

NOTE: All connections will be given as red/black combinations for the rest of the lab.

(8) Measure the voltage across R2 (DE). Record the data in Table 1.

(9) Measure the total voltage (AF). Record the data in Table 1.

(10) The current sensor needs to be in series with the circuit elements, whereas the voltage sensor is used in parallel. Therefore, we need to "rewire" the circuit slightly to get current readings. To measure the current across R1:

- Disconnect the metal tabs at C and D
- Remove the straight connector

- Reinsert the metal tabs
- Attach the red lead of the current sensor to the metal tab at C
- Attach the black lead of the current sensor to the metal tab at D

(11) Record your data in Table 1.

(12) Replace and reconnect the straight module.

(13) Repeat steps 10 & 11 to measure the current across R2 (except you will remove the module between E and F).

(14) Replace R2 with a 100 Ω resistor and repeat the experiment.

(15) Replace R1 with a 100 Ω resistor and repeat the experiment.

(16) Calculate R_{eq}, V_1+V_2, and $I_{circuit}$ from your data.

(17) Disconnect all sensors from the circuit. Turn off the current sensor.

Parallel Circuit

(18) Build a circuit with two 33 Ω resistors in parallel using the modular components. Don't forget to connect them with the tabs as well.

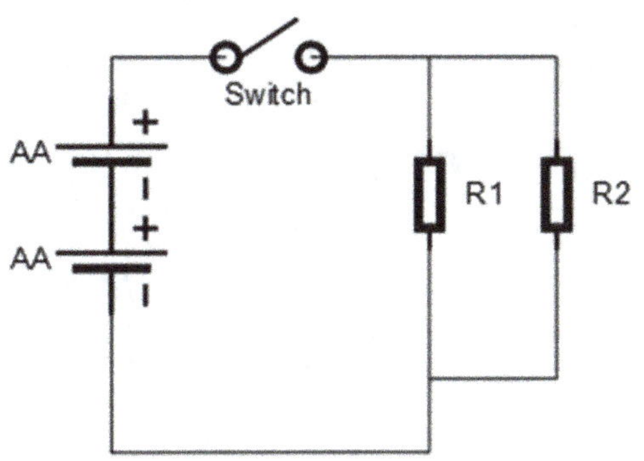

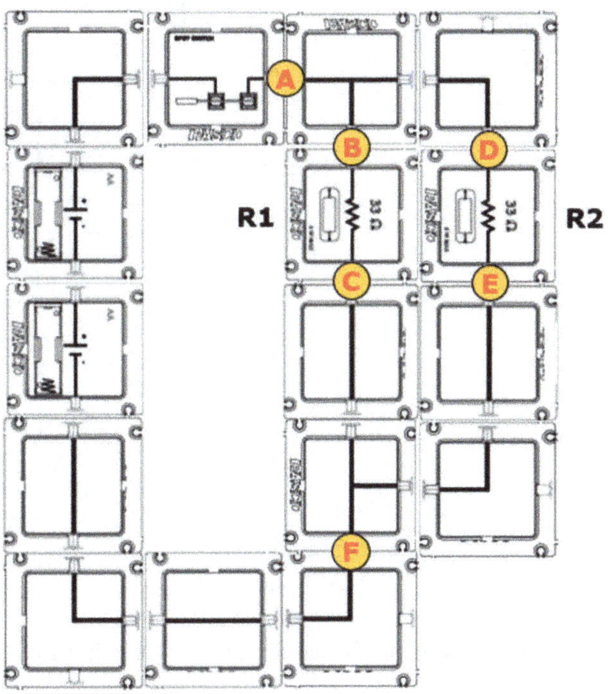

(19) Zero both sensors if needed and ensure that the switch is down.

(20) Measure the voltage across R1 (BC). Record the results in Table 2.

(21) Measure the voltage across R2 (DE). Record the results in Table 2.

(22) Measure the total voltage (AF). Record the results in Table 2.

(23) Using the procedure described earlier, measure the current through R1 by removing the straight module just below C. Record the results in Table 2.

(24) Replace and reconnect the straight module.

(25) Similarly, measure the current through R2 by removing the straight module just below E. Record the results in Table 2.

(26) Measure the circuit current by removing the straight module just below F. Record the results in Table 2.

(27) Replace R2 with a 100 Ω resistor and repeat the experiment.

(28) Replace R1 with a 100 Ω resistor and repeat the experiment.

(29) Calculate R_{eq} and I_1+I_2 from your data.

(30) Disconnect all sensors from the circuits.

Series-Parallel Circuit

(31) Build a circuit with two 33 Ω resistors in parallel and in series with a 100 Ω resistor using the modular components. Don't forget to connect them with the tabs as well.

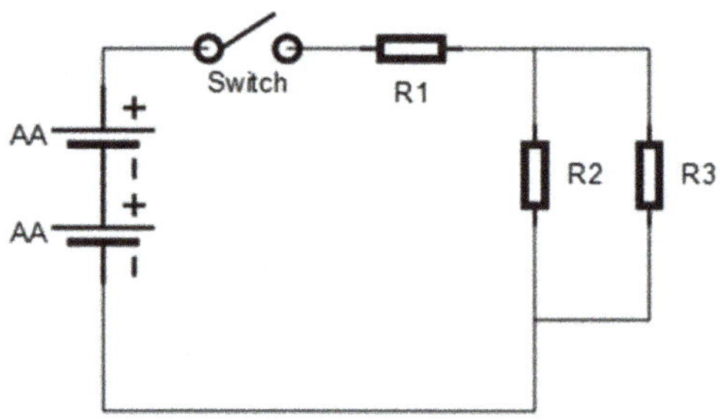

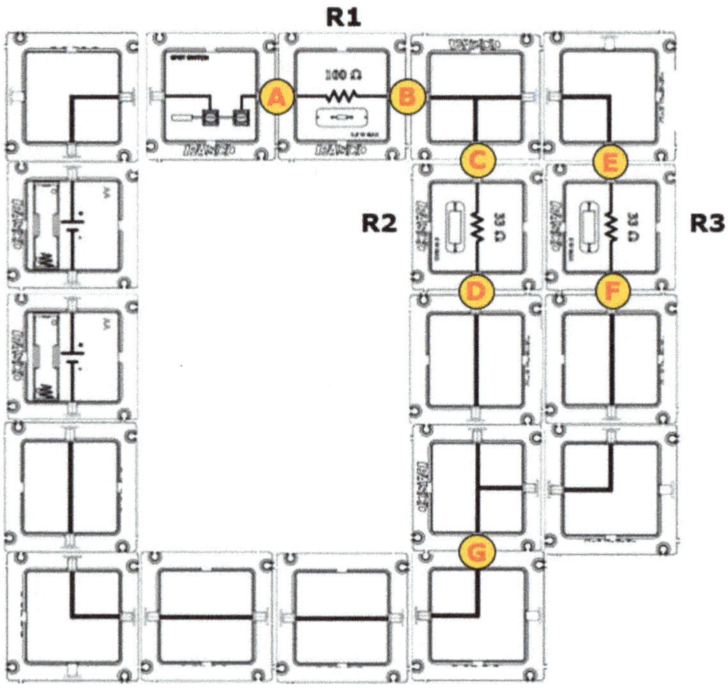

(32) Zero both sensors if needed and ensure that the switch is down.

(33) Measure the voltage across R1 (AB), R2 (CD), R3 (EF) and the total (AG). Record the results in Table 3.

(34) Using the procedure described earlier, measure the current through R1, R2, and R3.

(35) Calculate R_{eq} and $I_{circuit}$ from your data.

(36) Move the switch to the off position.

(37) Complete Questions 1–6.

RC CIRCUITS

BACKGROUND

An RC circuit is a circuit with both a resistor and a capacitor (C). A capacitor stores energy and a resistor placed in series with it will control the rate at which it charges or discharges. If a capacitor of capacitance C, initially charged to a potential V_0 is connected across a resistor R, a time-dependent current will flow according to Ohm's law. This time dependence turns out to be exponential in nature. The parameter that describes this time dependence is the time constant:

$$\tau = RC$$

Charging Capacitor

The charge on a charging capacitor as a function of time, and the current through the resistor as a function of time are given by:

$$q(t) = C\varepsilon\left(1 - e^{-\frac{t}{RC}}\right) = Q\left(1 - e^{-\frac{t}{\tau}}\right) \quad \text{and} \quad I(t) = I_0 e^{-\frac{t}{\tau}}$$

respectively. As the charge on the capacitor increases, the current through the resistor decreases, as does the voltage difference across the resistor.

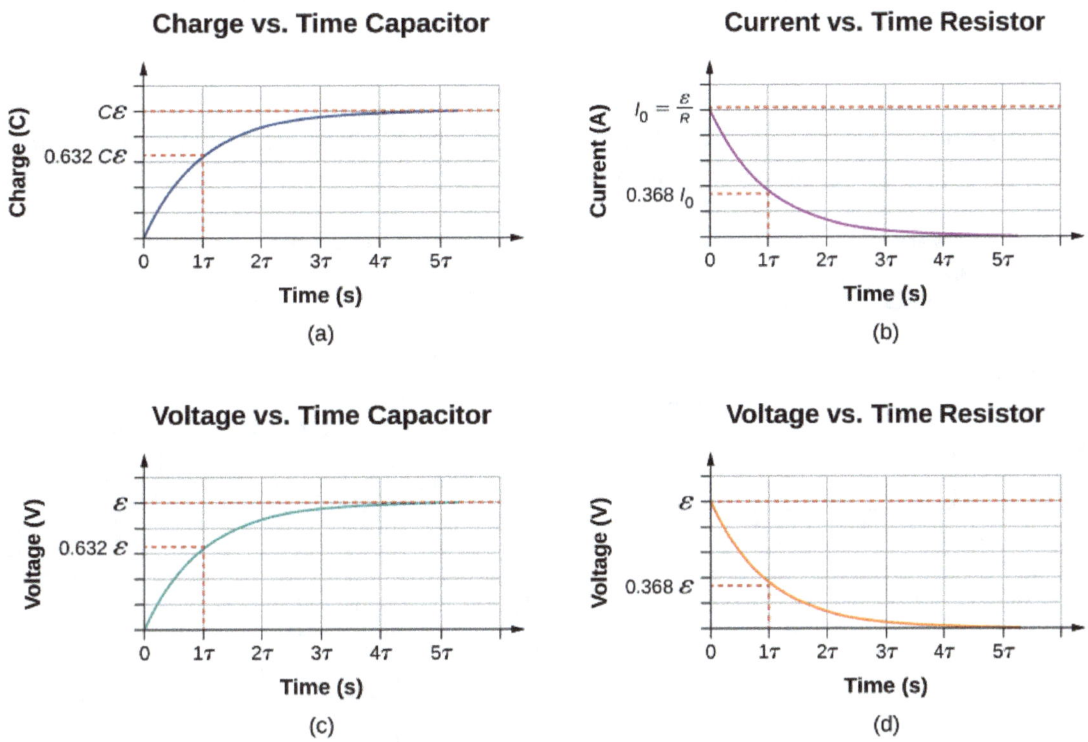

Discharging Capacitor

The charge on a discharging capacitor as a function of time, and the current through the resistor are:

$$q(t) = Qe^{-\frac{t}{\tau}} \quad \text{and} \quad I(t) = -\frac{Q}{RC}e^{-\frac{t}{\tau}} = -Ie^{-\frac{t}{\tau}}$$

respectively. Note the exponential decreasing functions.

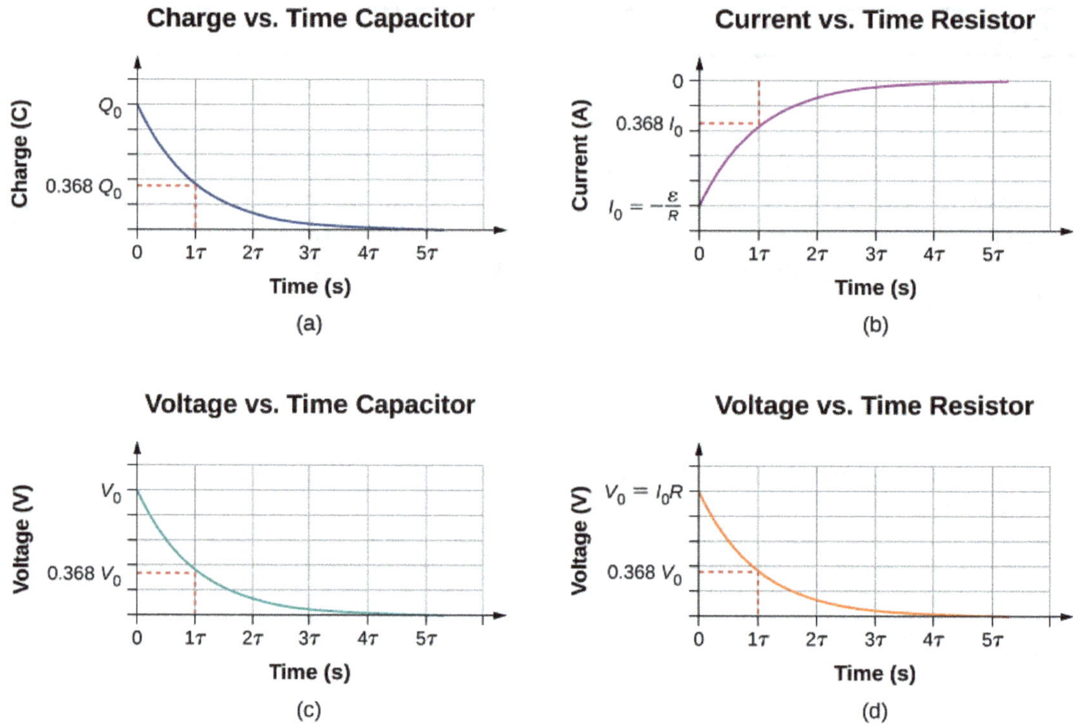

PROCEDURE

(1) Build a circuit with a SPDT switch, a 33 Ω resistor and a 0.22 F capacitor with the modular components. Don't forget to connect them with the tabs as well.

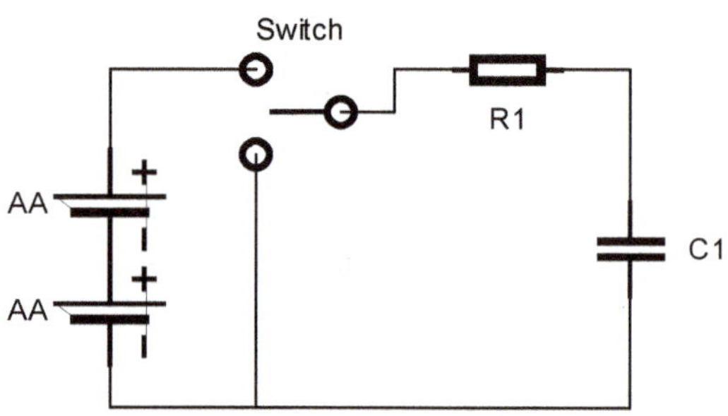

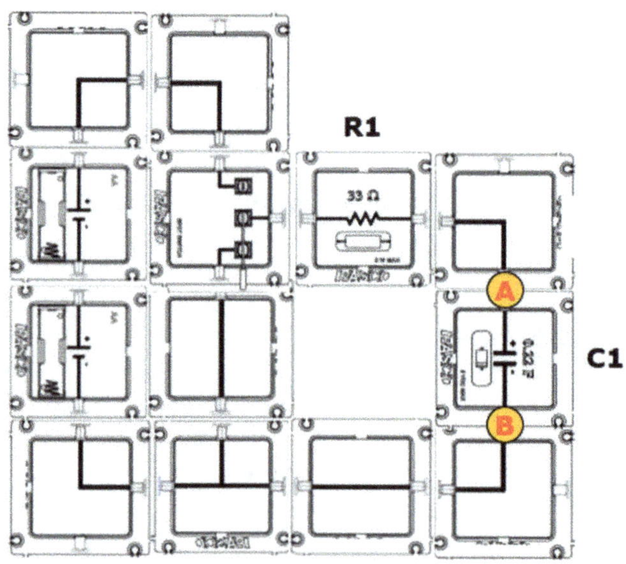

Changing Resistance in an RC Circuit

(2) Place the switch in the "discharge" position (towards the bottom of the figure). Wait 60 seconds for the capacitor to discharge, then place the switch in the center position (not connected to anything).

(3) The voltage sensor should still be on – if not, turn it on and connect to it.

(4) Configure the view for 1 graph (Potential vs Time).

(5) In the Data Collection Settings, change the End Collection to 60 s.

(6) Place the red lead of the voltage sensor at A and the black lead at B.

(7) Place the switch in the "charge" position (towards the top) and immediately click Collect to begin data collection.

(8) Once data collection stops, place the switch in the center position.

(9) Fit a natural exponential curve to the data and report the results in Table 4.

(10) Place the switch in the "discharge" position (towards the bottom) and immediately click Collect to begin data collection.

(11) Once data collection stops, place the switch in the center position.

(12) Fit a natural exponential curve to the data and report the results in Table 4.

(13) Switch out the 33 Ω resistor for the 100 Ω.

(14) Repeat steps 7–12 with the new resistor. Report the results in Table 4.

(15) Place the switch in the center position.

Changing Capacitance in an RC Circuit

(16) We only have 0.22 F capacitors, but we can change the capacitance by using multiple 0.22 F capacitors in series or in parallel.

(17) Build a circuit with a SPDT switch, a 33 Ω resistor and two 0.22 F capacitors in series with the modular components. Don't forget to connect them with the tabs as well.

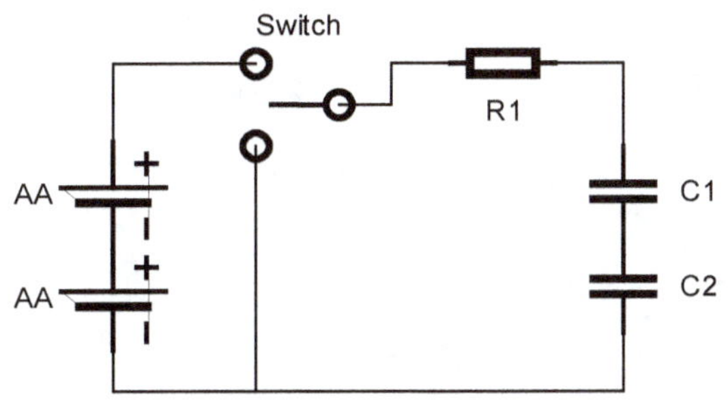

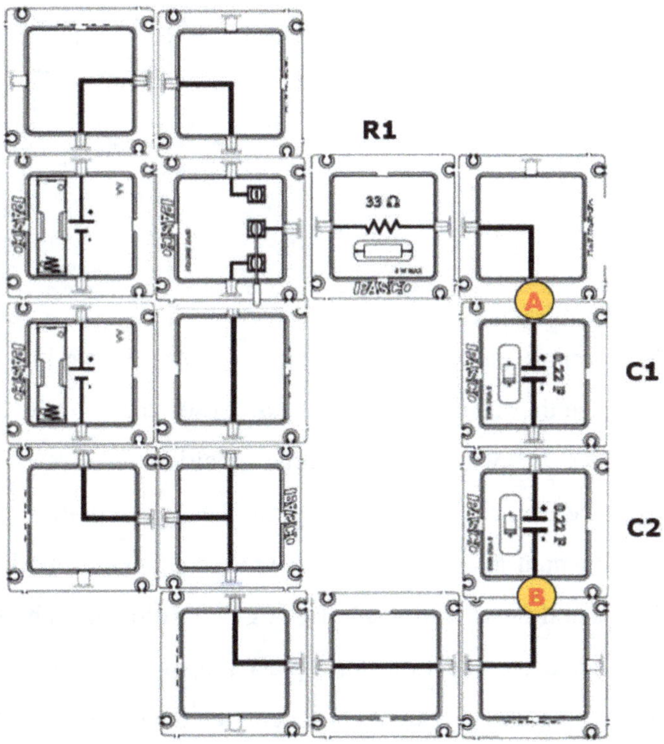

(18) Compute the equivalent capacitance and enter it in Table 4 (Trial 3 capacitance).

(19) Place the red lead of the voltage sensor at A and the black lead at B.

(20) Place the switch in the "charge" position and immediately click Collect to begin data collection. Once data collection stops, place the switch in the center position.

(21) Fit a natural exponential curve to the data and report the results in Table 4.

(22) Place the switch in the "discharge" position and immediately click Collect to begin data collection.

(23) Once data collection stops, place the switch in the center position.

(24) Fit a natural exponential curve to the data and report the results in Table 4.

(25) Build a circuit with a SPDT switch, a 33 Ω resistor and two 0.22 F capacitors in parallel with the modular components. Don't forget to connect them with the tabs as well.

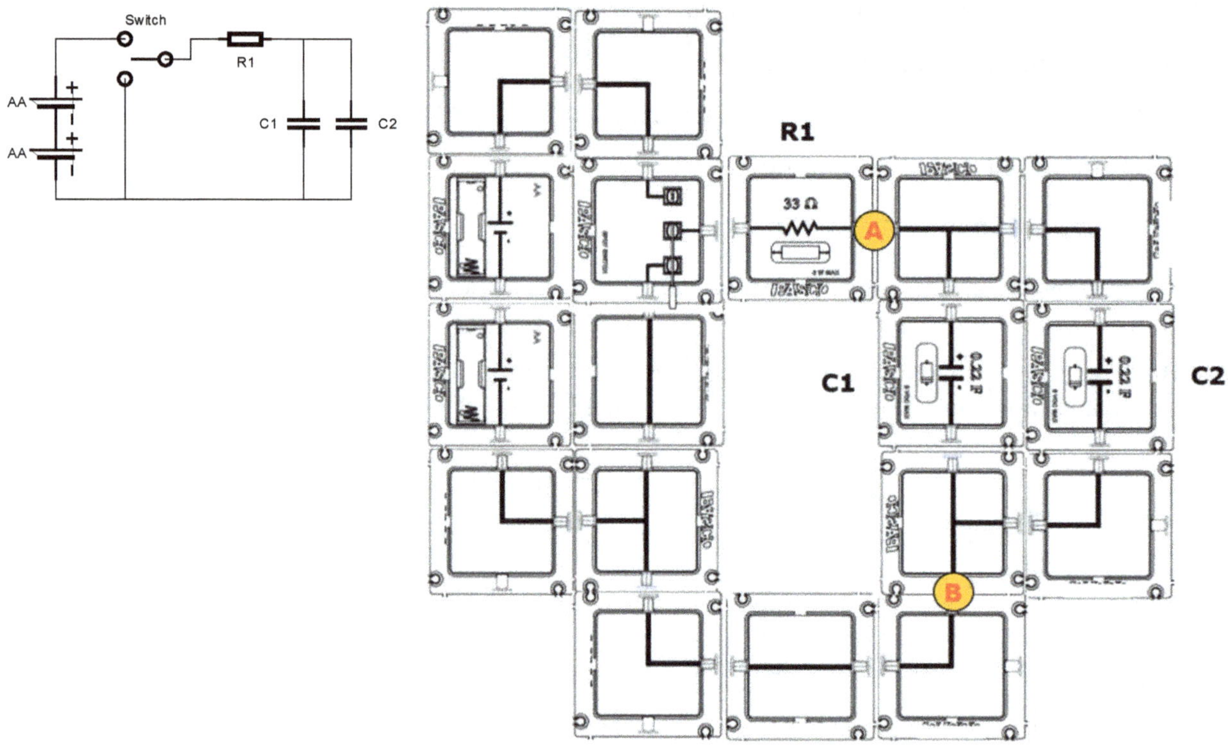

(26) Compute the equivalent capacitance and enter it in Table 4 (Trial 4 capacitance).

(27) Place the red lead of the voltage sensor at A and the black lead at B.

(28) Place the switch in the "charge" position and immediately click Collect to begin data collection.

(29) Once data collection stops, place the switch in the center position.

(30) Fit a natural exponential curve to the data and report the results in Table 4.

(31) Place the switch in the "discharge" position and immediately click Collect to begin data collection.

(32) Once data collection stops, place the switch in the center position.

(33) Fit a natural exponential curve to the data and report the results in Table 4.

(34) Complete Questions 7–13.

LAB 19 – ASSIGNMENT

QUESTIONS – RESISTORS IN SERIES AND PARALLEL

Table 1

	Resistance (Ω)			Voltage (V)				Current (A)		
Combo	**R_1**	**R_2**	**R_{eq}**	**V_1**	**V_2**	**V_1+V_2**	**V_{total}**	**I_1**	**I_2**	**I_{circ}****
1	33	33								
2	33	100								
3	100	100								

*V_{total} is measured ** $I_{circuit} = V_{total}/R_{eq}$

Table 2

	Resistance (Ω)			Voltage (V)			Current (A)			
Combo	**R_1**	**R_2**	**R_{eq}**	**V_1**	**V_2**	**V_{total}**	**I_1**	**I_2**	**I_1+I_2**	**I_{circ}****
1	33	33								
2	33	100								
3	100	100								

*V_{total} is measured ** $I_{circuit} = V_{total}/R_{eq}$

Table 3

Resistance (Ω)				Voltage (V)				Current (A)			
R_1	**R_2**	**R_3**	**R_{eq}**	**V_1**	**V_2**	**V_3**	**V_{total}**	**I_1**	**I_2**	**I_3**	**I_{circ}****
100	33	33									

*V_{total} is measured ** $I_{circ} = V_{total}/R_{eq}$

(1) For resistors in a series circuit, what is the relationship between V_1, V_2 and V_{total}?

(2) For resistors in a series circuit, what is the relationship between I_1, I_2 and $I_{circuit}$?

(3) For resistors in a series circuit, voltage through the 100 Ω resistor is more than it is through the 33 Ω resistor. Why?

(4) For resistors in a parallel circuit, what is the relationship between V_1, V_2 and V_{total}?

(5) For resistors in a parallel circuit, what is the relationship between I_1, I_2 and $I_{circuit}$?

(6) For the series-parallel circuit below, calculate the equivalent resistance if: R1=10 Ω, R2=12 Ω, R3=4 Ω, R4=1 Ω, R5=10 Ω, R6=8 Ω, and R7=2 Ω.

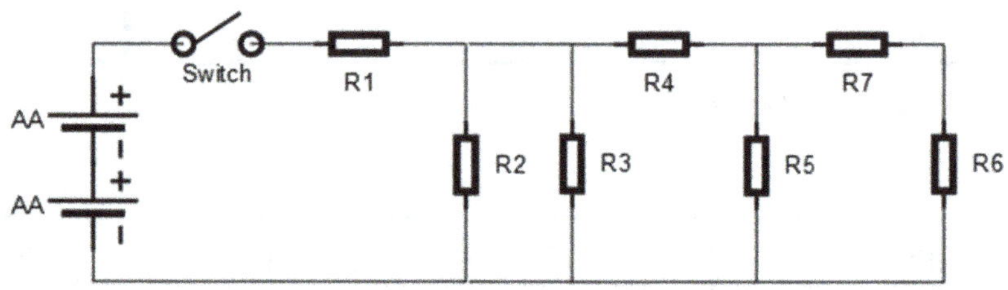

QUESTIONS – RC CIRCUITS

Table 4

			Charging				Discharging			
Trial	**R**	**C**	**A**	**B**	**C**	**RMSE**	**A**	**B**	**C**	**RMSE**
1	33 Ω	0.22 F								
2	100 Ω	0.22 F								
3	33 Ω	0.11 F								
4	33 Ω	0.44 F								

(7) Calculate the time constant for each trial.

(8) Which curve fit parameter does 1/RC (the inverse time constant) roughly equal (capacitor values are approximate, so the 0.22 F can be as much as 20% off).

(9) What was the effect on the time constant if we increase the value of the resistor?

(10) Does increasing the resistance cause the capacitor to charge & discharge more quickly or more slowly?

(11) What is the effect on the time constant if we increase the value of the capacitance?

(12) Show that the unit for the time constant is seconds (i.e., show that ohm x farad = seconds).

(13) Suppose an RC circuit has a time constant of 5 s. The voltage as a function of time for the charging of the capacitor is given by: $Q\left(1-e^{-\frac{t}{\tau}}\right)$ and the voltage for the discharging capacitor is $Qe^{-\frac{t}{\tau}}$. At what time, t, are these equal?

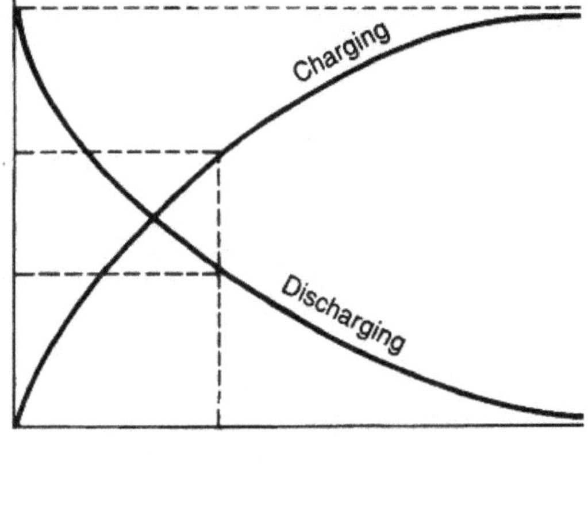

LAB 20 | ELECTROMAGNETIC ENERGY

GETTING STARTED

OVERVIEW

In this lab, you will explore the nature of electromagnetic radiation and topics that we did not necessarily discuss in class. You will also experimentally determine the value of Planck's Constant.

EQUIPMENT

- Lab Computer
- Planck's Constant Apparatus
- Multimeter
- Patch Cables

EM WAVES AND ENERGY

BACKGROUND

Electromagnetic (EM) energy travels through space as a transverse wave, which consists of oscillating electric (E-field) and magnetic fields (B-Field). EM energy can be represented by the familiar sine curve, if we consider the E-field and the B-field independently.

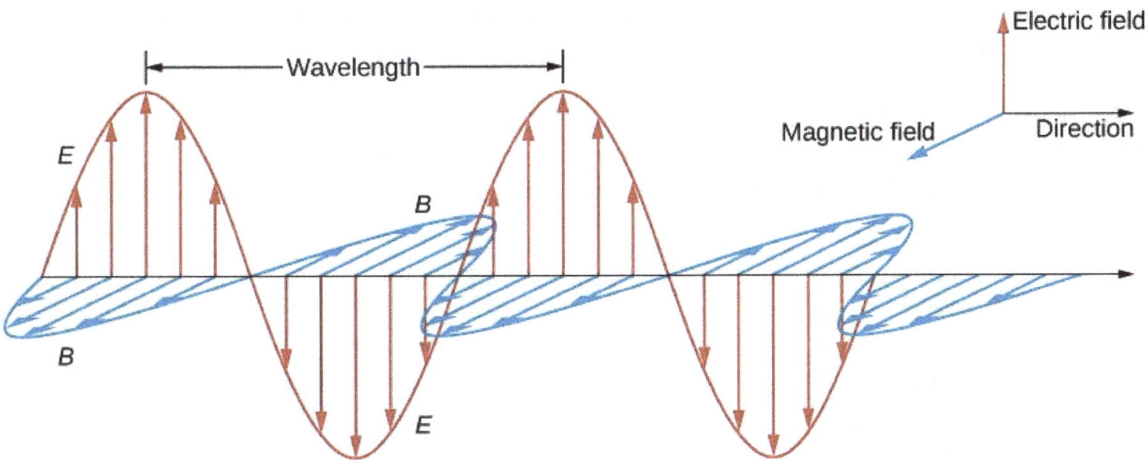

Wavelength is the distance between successive equivalent points in the wave. Distance normally depends on speed; however, all forms of EM energy travels at the same speed – the speed of light ($3 \cdot 10^8$ m/s). Thus, the base unit for wavelength is meters, but it is often converted to other units. Another characteristic of EM energy is frequency, which is how many complete cycles (one complete cycle in per wavelength) pass a point in 1 second. Frequency is measured in Hertz (Hz).

The relationship between frequency, wavelength, and the speed of light is specific and can be rearranged to solve for any of the components. Since the speed of light is constant, if we know either wavelength (λ) or frequency (ν), we can calculate the other:

$$c = \frac{\lambda}{\nu}$$

Radiation Laws – EM energy is absorbed and emitted in discreet amounts (quanta) that is wavelength-dependent. So, if we know the wavelength (or the frequency), we can calculate the amount of this energy using Plank's Equation. Furthermore, all objects warmer than absolute zero constantly emit EM energy; the amount depends on the object's surface temperature and a property called emissivity.

If we know the amount of radiation emitted by an object, we can calculate its temperature using the Stefan-Boltzman Law. Conversely, if we know an object's temperature, we can calculate the amount of radiation it emits. Finally, we can relate an object's temperature to the wavelength with the maximum amount of emitted energy using Wien's Displacement Law. These equations are summarized below along with values for the constants.

Light Formula	Plank's Equation	Stefan-Boltzman Law	Wien's Law
$c = \lambda \nu$	$Q = \frac{hc}{\lambda}$	$W = \varepsilon \sigma T^4$	$\lambda_{max} = \frac{2897.8\ \mu m \cdot K}{T}$
	Speed of EM energy	$c = 3 \cdot 10^8\ m \cdot s^{-1}$	
	Plank's Constant	$h = 6.626 \cdot 10^{-34}\ m^2 \cdot kg \cdot s^{-1}$	
	Stefan-Boltzman Constant	$\sigma = 5.67 \cdot 10^{-8}$ watts $\cdot\ m^{-2} \cdot K^{-4}$	
	emissivity	ε = depends on material and will be given to you	

The light formula and Plank's Equation primarily concern transmitted or reflected energy. On the other hand, the Stefan-Boltzman Law and Wien's Law are more useful for emitted energy and anything to do with temperature. For an object to be visible due to emitted energy alone, it must be very, very hot – a few thousand kelvins or more. Note that this is not the same as same as being visible due to reflected EM energy, which is our everyday experience.

The chart plots the relationship between intensity and wavelength. Note that an object emitting in the red portion of the spectrum is cooler than something emitting in the blue spectrum (the expression "red hot" is somewhat of a misnomer). The wavelength with the most intensity is λ_{max} and it decreases with increasing temperature.

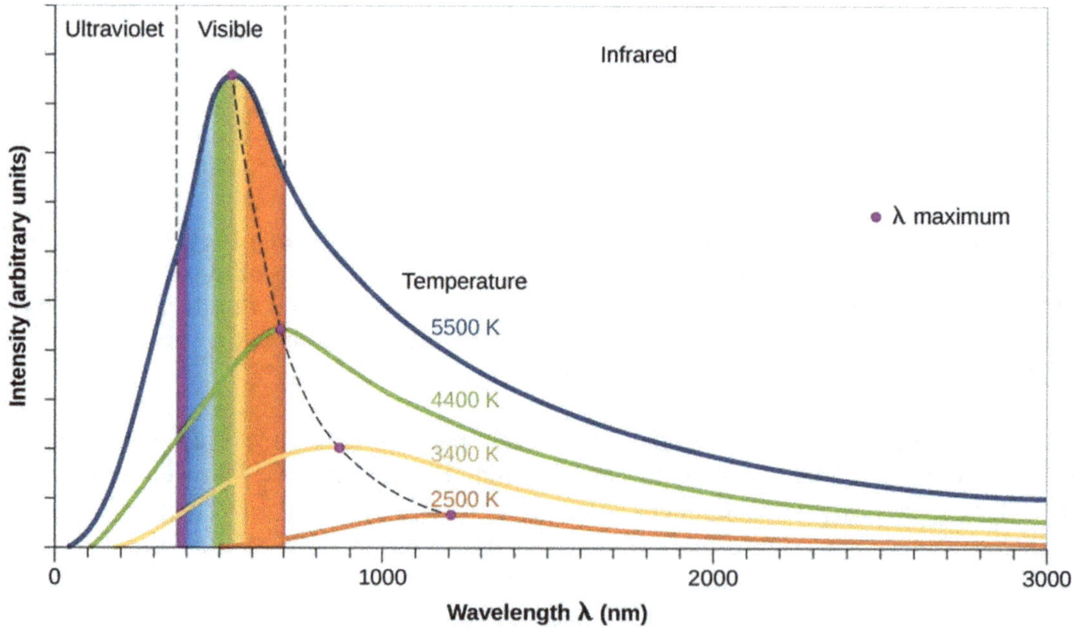

The Electromagnetic Spectrum – EM energy varies in frequency/wavelength along an almost infinite scale. Due to the inverse relationship between frequency and wavelength, we find that as frequency increases, the wavelength decrease, and vice-versa. Energy has an inverse relationship with wavelength (Plank's Equation), which means it has a direct relationship with frequency. So, as frequency increases, so does energy. These general relationships are summarized in the figure below.

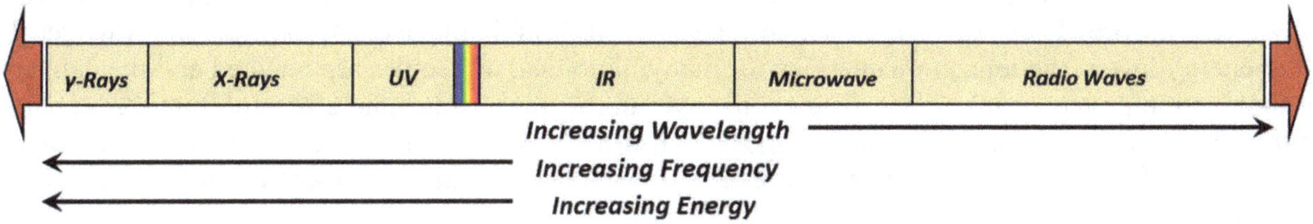

The portions of the EM spectrum we are more likely to experience are centered around visible light and consists of a portion of the ultraviolet (UV), visible, and a portion of the infrared (IR). Focusing on these portions of the spectrum, the EM spectrum is divided according to the following table.

Spectrum	Sub-division	Limits (λ)
Ultraviolet (UV)	Near UV (UV-A)	320 – 400 nm
Visible (VIS)	Blue (B)	400 – 500 nm
	Green (G)	500 – 600 nm
	Red (R)	600 – 700 nm
Infrared (IR)	Near IR (NIR)	700 – 1300 nm
	Mid IR (MIR)	1300 – 3000 nm

PROCEDURE

(1) Complete Questions 1–7.

EM RADIATION VS MATTER

BACKGROUND

When EM radiation interacts with an object, it can be (1) absorbed by the object, (2) transmitted through the object, or (3) reflected from the object. Which of these occurs is a function of wavelength/ frequency.

Microwave Interactions – Microwave photons interact with non-metallic materials by causing molecules to rotate, which generates heat through increased molecular motion. In contrast, metallic materials absorb microwaves. Most substances, including the human body, are largely transparent to microwaves. At high intensities, e.g., a microwave oven, the waves repeatedly pass through the material, resulting in heating via molecular rotation and twisting.

Infrared Interactions – Infrared photons interact with non-metallic materials by causing molecules to rotate, vibrate, or bend, which generates heat through increased molecular motion. Infrared energy is absorbed more strongly by objects than microwaves. The result of infrared absorption in biological tissues is heating of the tissue. Most non-metallic materials absorb infrared radiation while metals tend to reflect infrared energy.

Visible Light Interactions – Visible light photons interact with matter in the ways that we examined in optics: absorption, transmission, scattering, reflection, and refraction. The primary mechanism for the absorption of visible light is to elevate electrons to higher energy levels. There are many available states, so visible light is absorbed strongly. For example, red light can be transmitted through the hand or a fold of skin with a strong light source, showing that the red end of the spectrum is not absorbed as strongly as the violet end.

Ultraviolet Interactions – Ultraviolet photons interact with matter primarily through absorption, causing electronic excitation, the breaking of chemical bonds, and ionization due to their high energy. The near ultraviolet is absorbed very strongly in the surface layer of the skin by electron transitions. As we reach higher energy levels, the ionization energies for many biological molecules are reached and the dangerous photoionization processes take place. The ozone layer in the upper atmosphere is important for human health because it absorbs most (but not all) of the harmful ultraviolet radiation from the sun before it reaches the surface. Sunburn is primarily an effect of UV, and ionization produces the risk of skin cancer.

X-Ray Interactions – The energies of x-ray photons are too high to be absorbed by just moving electrons between energy levels; so, they are not absorbed. Instead, X-rays interact by ejecting electrons from the atom entirely, making x-rays a form of ionizing radiation.

PROCEDURE

(1) Complete Questions 8–9.

PLANCK'S CONSTANT

BACKGROUND

In this course, we've studied energy from a classical view in which energy is continuous and can take on any value. In quantum physics, energy is in specific amounts, known as quanta. Planck's constant is a number that relates a photon's energy to its frequency. Planck's constant was formulated as part of an effort to produce a mathematical expression that accurately predicted the observed spectral distribution of thermal radiation from a blackbody.

Planck's constant can be expressed in two ways depending on the units:

- $6.626 \cdot 10^{-34}$ J·s
- $4.136 \cdot 10^{-15}$ eV·s

Experimental Notes – Even though the value of Planck's constant is extraordinarily small, we have a clever method of determining this value using LEDs (light emitting diodes). LEDs are made from two types of "doped" semiconductor materials joined together. One type has an excess of electrons (n-type) and the other a lack of electrons (p-type). When an electrical current passes through this p-n junction, energy in the form of photons is released. The color of the light emitted from an LED is determined by the energy of the photons, which can be custom-made by changing the chemical composition of the semiconductor materials. We use LEDs because each color of LED has a different threshold voltage at which electrons start being produced. Measuring this voltage, along with known values for the emission wavelengths, provides a mechanism to find a value for Planck's constant.

PROCEDURE

(1) Set up the Planck's Constant apparatus. It should be plugged into an outlet and the yellow patch cord should connect the two yellow jacks. Connect a multimeter to the apparatus (red to the first yellow jack, and black to the black jack on the left).

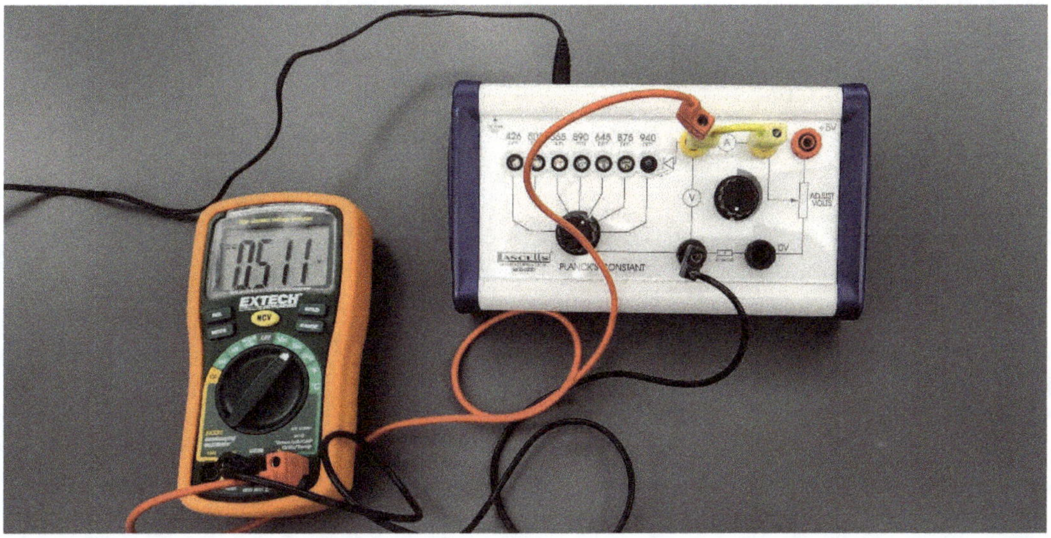

(2) Ensure the voltage knob on the apparatus is fully CW. This is the off position, which is backwards from most knobs in which full CW is full ON.

(3) Turn the selector on the multimeter to V-DC so that it will measure DC voltage.

(4) Select the first LED (426 nm) by changing the wavelength selector.

(5) Slowly increase the voltage by turning the voltage know CCW until the LED just begins to light up. This will produce a faint glow which will be easier to see if you use the black tube with the eyepiece.

(6) Record the voltage from the multimeter in Table 1.

(7) Turn the voltage back to zero.

(8) Select the next wavelength and repeat the process.

(9) Continue until you have done the first 5 wavelengths and Table 1 is completed.

(10) Open Graphical Analysis and select manual entry.

(11) Enter the data, using frequency (Hz not nm) as the *x*-values and the voltage measurements as the *y*-values. [**NOTE** – to enter your x-values in scientific notation, you'll need to configure the *x*-column. Click on the three dots for the *x*-column. In the dialog box, check "Use Scientific Notation" and then change Displayed Precision to say 3 decimal places].

(12) Apply a linear curve fit.

(13) Complete Questions 10–15.

LAB 20 – ASSIGNMENT

QUESTIONS – EM WAVES AND ENERGY

(1) Fill in the missing values (the desired units are given) in the table by using the relationship between wavelength, frequency, and the speed of light. Give the answers to 2 decimals. Show your work.

Frequency	Wavelength
422 **MHz**	**m**
THz	35 **μm**

Use the graph for Q2–5.

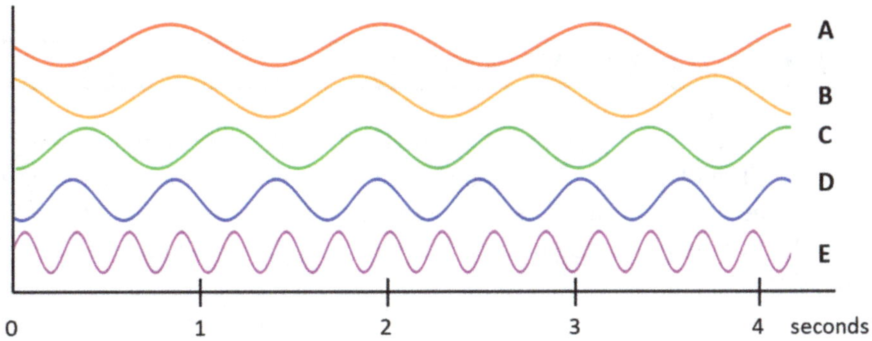

(2) Which line has the highest frequency?

(3) Which line has the longest wavelength?

(4) What is the frequency of line B?

(5) Fill in the missing values in the table by using the radiation laws and the table for the divisions of the EM spectrum. Be careful with units. Give the answers to 2 decimals.

Spectrum Subdivision	Wavelength (nm)	Frequency (Hz)	Energy ($m^2 \cdot kg \cdot s^{-2}$)
UV-A	370		$5.37 \cdot 10^{-19}$
	460		
		$2.73 \cdot 10^{14}$	

(6) How much energy is emitted by dry soil ($\varepsilon = 0.92$) that is at a temperature of 20°C.

(7) If a bright star burns at 7000 K, what color is the star?

QUESTIONS – EM RADIATION VS MATTER

(8) When light interacts with objects that are much smaller than the wavelength, the light is scattered rather than reflected. The electrons of the "small object" are shaken up and down at the same time by the electric field of the light wave and therefore radiate that frequency of light in all directions. It turns out that the higher the frequency of light, the more the light is scattered. Use this information, explain why the sky is blue.

(9) Use the information in Q8 and this diagram to explain why sunsets are red.

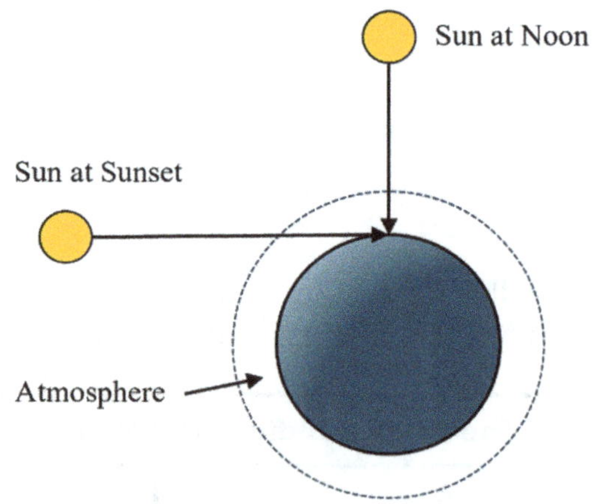

QUESTIONS – PLANCK'S CONSTANT

Table 1

Wavelength (nm)	Color	Frequency (Hz)	Threshold Voltage
426 nm			
501 nm			
565 nm			
590 nm			
645 nm			

(10) What was the correlation coefficient (r) for your linear fit?

(11) What was the slope from your linear fit?

(12) The slope from your linear fit is Planck's constant expressed in eV units ($h = 4.136 \cdot 10^{-15}$ eV·s).What is your percent error?

(13) We did not use the 875 and 940 nm LEDs. Why not?

(14) A red laser pointer emits light with a wavelength of 700nm and a green one emits light at 500nm. Find the amount of energy per photon for each laser pointer.

(15) UVA radiation have a wavelength of about 360 nm, while UVB have a wavelength of 300 nm. UVA radiation penetrates more deeply into the skin, but UVB more readily causes cellular damage via sunburn. How much more energy does UVB have compared to UVA?

LAB 21 | ATOMIC SPECTRA

GETTING STARTED

OVERVIEW

In this lab, you will explore the relationship between observed wavelengths in the Hydrogen spectrum and the energies involved when electrons undergo transitions between energy levels.

EQUIPMENT

- Lab Computer
- SpectroVis Plus
- Spectrophotometer Optical Fiber
- Spectrum Tube Power Supply
- Gas Discharge Tubes

HYDROGEN SPECTRUM

BACKGROUND

In the late 19th century, physicists could not explain why excited gases emitted light only at specific wavelengths. In 1885, Johann Balmer, discovered an empirical formula that described the wavelengths of visible spectral lines in hydrogen, where m is an integer greater than 2.

$$\lambda = 364.56\text{ nm}\left(\frac{m^2}{m^2 - 2^2}\right)$$

In 1888, Johannes Rydberg reorganized Balmer's equation into a more general form still used today:

$$\frac{1}{\lambda} = R\left(\frac{1}{n_1^2} - \frac{1}{n_2^2}\right)$$

Here, R is the Rydberg constant ($R = 1.097 \cdot 10^7$ /m), and n_1 and n_2 are integers such that $n_1 < n_2$. This equation also helped identify other series of spectral lines, including the Lyman series (ultraviolet) and the Paschen series (infrared).

We can invert $\frac{1}{\lambda}$ to get the wavelength of the emitted line and use $c = \lambda f$ to find the frequency.

In 1911, Niels Bohr explained the origin of spectral lines by proposing that electrons move in specific orbits, each corresponding to a particular energy level. In this model, an electron can only exist in one of these discrete energy states, and it normally remains in the lowest energy level (ground state) and it does not emit light while in this condition. However, if an atom absorbs enough energy through a collision or by absorbing a photon, the electron can move from the ground state ($n = 1$) to an excited state ($n > 1$) which has higher energy. When the electron returns to the lower energy level, it releases the excess energy in the form of a photon. The spectral lines correspond to these transitions as electrons move from higher to lower energy states.

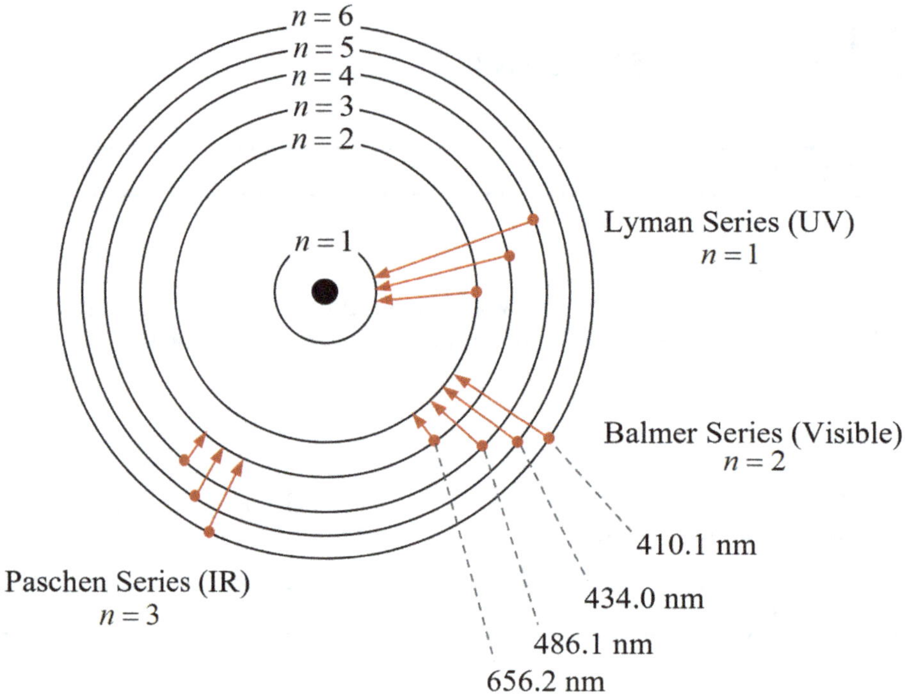

The energy of the light emitted can be calculated using Planck's equation:

$$E = hf$$

where, h is Planck's constant ($h = 6.63 \cdot 10^{-34}$ J·s) and f is frequency. Finally, we can combine Rydberg's equation and Planck's equation to eliminate frequency and express energy only in terms of n_1 and n_2.

$$E = 2.18 \cdot 10^{-18}\ \text{J} \left(\frac{1}{n_1^2} - \frac{1}{n_2^2} \right)$$

Hydrogen is a good element to examine spectroscopically as it is the simplest atom. If an electric current is passed through a glass tube that contains hydrogen gas at low pressure, the tube gives off blue light. When we examine this light with a spectrophotometer, there are 4 primary emission lines: 656.2, 486.1, 434.0, and 410.1 nm. There are others, but most of these are not energetic (particularly in the blue region) and it is difficult to get good peaks experimentally without very expensive equipment.

Instead, we'll use the reference gas discharge spectra for Hydrogen. You'll note that there are four main peaks here (though one is quite small). These correspond to the values given earlier: 656.2, 486.1, 434.0, and 410.1 nm (see figure above).

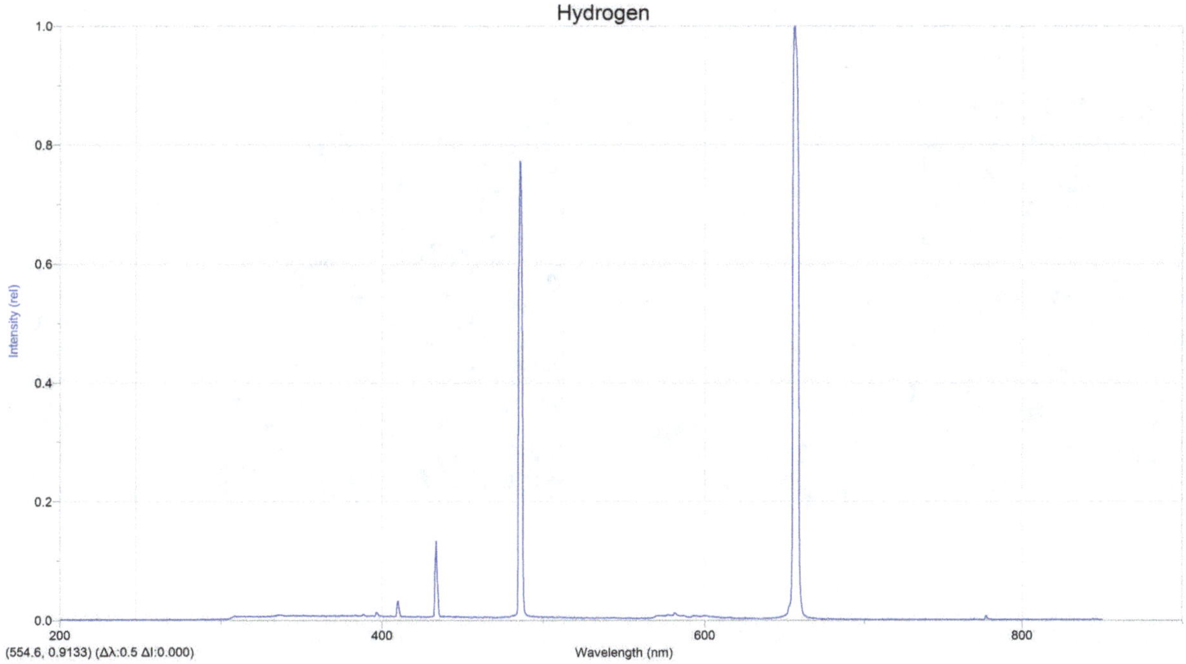

PROCEDURE

(1) Complete Questions 1–3.

OTHER SPECTRA

BACKGROUND

As discussed earlier, one way to excite an electron is to pass an electric current through an confined gas at low pressure. Since the electron energy levels are unique for each element, every gas discharge tube will glow with a distinctive color, depending on the identity of the gas. "Neon" signs are familiar examples of gas discharge tubes. However, only signs that glow with a red-orange color are filled with neon. Signs of other colors contain different gases or mixtures of gases.

PROCEDURE

(1) Insert the Optical Fiber into the spectrophotometer so that the cuvette-insert fits into the opening.

(2) Turn on the power to the spectrophotometer (on the side).

(3) Start the Spectral Analysis application (NOT Graphical Analysis). Select your sensor from the list of Discovered Wireless Devices. Alternatively, you can connect the spectrometer to the computer directly.

(4) Select Emissions – vs Wavelength (Full Spectrum).

(5) Click on Collection Settings (the gear icon in the upper right).

(6) Change the integration time to 30 ms (it may already be set to this value). Click somewhere outside of the Settings box to close it.

(7) Insert the fiber optical cable into the spectrometer and the attachment on the back of the spectrum tube power supply.

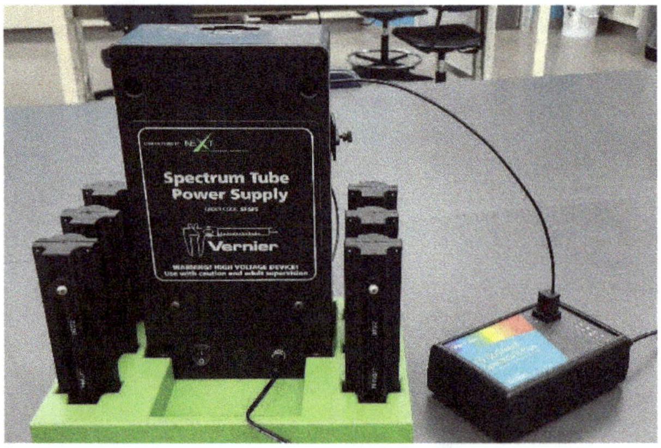

(8) Insert the Argon tube into the power supply.

(9) Turn on the power – this energizes the gas in the tube and it should begin to glow.

CAUTION

- Do not touch the gas discharge tubes when they are energized (glowing). They get very hot.

(10) Click Collect to begin data acquisition.

(11) Click stop after approximately 1 second.

(12) Determine the intensity and wavelengths of the 6 largest peaks (click on them) and enter the data in Table 1. Place them in order with the one with the largest intensity first.

(13) Repeat the process for Helium and Neon.

(14) Fill in Table 1 with the appropriate data.

(15) Repeat for CO2 and Nitrogen. Do not fill data into Table 1 for either of these.

(16) Turn off the spectrophotometer.

(17) Complete Questions 4–6.

LAB 21 - ASSIGNMENT

QUESTIONS - HYDROGEN SPECTRUM

(1) Fill in the values for frequency and energy.

Peak	Wavelength (nm)	Frequency (Hz)	Energy (J)
1			
2			
3			
4			

(2) Use $E = 2.18 \cdot 10^{-18}\text{ J}\left(\frac{1}{n_1^2} - \frac{1}{n_2^2}\right)$, to determine values of n_1 and n_2 that give the light energies that approximate the values calculated in the table above. The best way to approach this is isolated n_1 and n_2 by dividing E by $2.18 \cdot 10^{-18}$, then match this value by substituting various pairs of integers in the $\left(\frac{1}{n_1^2} - \frac{1}{n_2^2}\right)$ part of the equation. Recall that $n_1 < n_2$ and that they are integers. Use 4 decimals for the ratio.

Peak	Energy (J) from above	Ratio (Energy/$2.18 \cdot 10^{-18}$)	n_1 and n_2
1			
2			
3			
4			

(3) Use $E = hf$ and $c = \lambda f$ to show that $\frac{1}{\lambda} = R\left(\frac{1}{n_1^2} - \frac{1}{n_2^2}\right)$ is equivalent to $E = 2.18 \cdot 10^{-18}\text{ J}\left(\frac{1}{n_1^2} - \frac{1}{n_2^2}\right)$.

QUESTIONS – OTHER SPECTRA

Table 1

	Argon		Helium		Neon	
Peak	**Wavelength**	**Intensity**	**Wavelength**	**Intensity**	**Wavelength**	**Intensity**
1						
2						
3						
4						
5						
6						

(4) Compared to hydrogen and helium, the spectrum for nitrogen is quite "messy". Why?

(5) What color was the CO_2 ionized gas? Based on the spectrum for CO_2.

(6) The five largest peaks in the sun's spectrum are labeled here. Based on this plot, what two elements make up most of the sun? Explain.

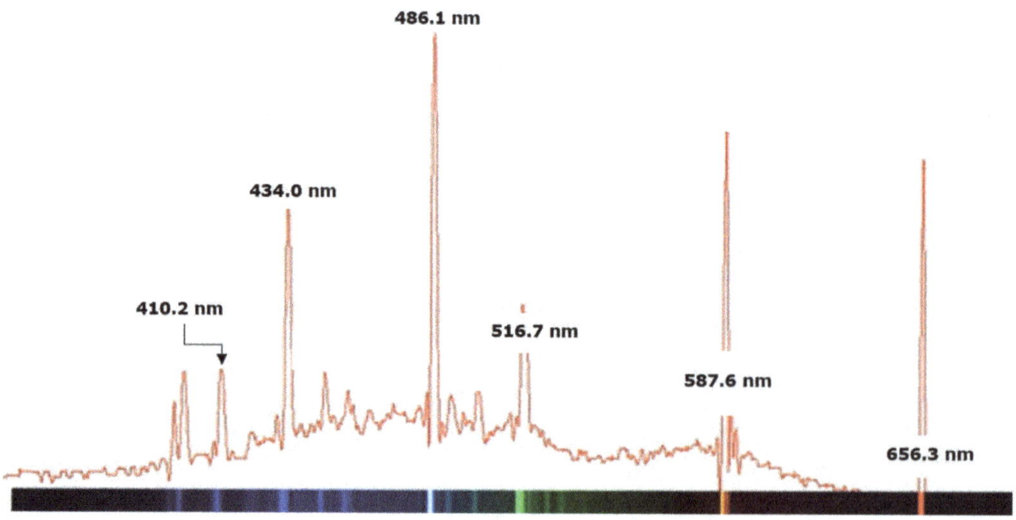

LAB 22 | RADIOACTIVE DECAY & HALF-LIFE

GETTING STARTED

OVERVIEW

In this lab, you will explore radioactive decay.

RADIOACTIVE DECAY

BACKGROUND

Radioactive (nuclear) decay occurs when an unstable atom loses energy by emitting radiation to reach a more stable energy state. Radioactive decay is an inherently random process in which the unstable atom's nucleus spontaneously decomposes to form a different nucleus, giving off radiation in the form of atomic particles or high energy rays in the process. This decomposition occurs at a constant, predictable rate that is referred to as half-life. A stable nucleus will not undergo this kind of decay and is non-radioactive.

There are three fundamental modes of radioactive decay: alpha, beta, and gamma. Alpha decay describes a process in which an unstable nucleus emits an alpha particle (two protons and two neutrons, equivalent to a He nucleus) to achieve stability. Beta decay describes the process in which an unstable nucleus emits a beta particle (an electron or proton) to achieve stability. Gamma decay describes the process in which an excited nucleus releases excess energy by emitting a high-energy photon (gamma ray) without changing its proton or neutron number. Gamma decay commonly follows immediately after alpha or beta decay.

Radioactive decay results in a redistribution of the basic nuclear particles, so we need a notation system that keeps track of where these particles are before and after a nuclear transformation.

Name	Symbol	Description
Parent Species	${}^{A}_{Z}X$	General symbol for any element X with atomic number Z & mass A
Daughter Species	${}^{A}_{Z}X'$	General symbol for any element X' with atomic number Z & mass A
Alpha Particle	${}^{4}_{2}\mathrm{He}$	High energy Helium nucleus (2 neutrons and 2 protons).
Beta Particle	${}^{0}_{-1}\mathrm{e}$	High energy electron.
Positron	${}^{0}_{1}\mathrm{e}$	Positron (particle with same mass as electron but positive charge)
Proton	${}^{1}_{1}\mathrm{p}$	Proton (nucleus of a Hydrogen atom)
Neutron	${}^{1}_{0}\mathrm{n}$	Neutron (particle with same mass as proton but neutral charge)
Gamma Ray	γ	Very-high energy EM radiation

To balance nuclear reactions, two rules must be followed:

- Mass number is conserved – the sum of the mass numbers (top numbers) before the decay must equal the sum of the mass numbers after the decay.
- Electric Charge is conserved – the sum of the atomic numbers (bottom) before the decay must equal the sum of the atomic numbers after the decay.

For example, the radioactive decay of uranium-238 to thorium-234 is given by:

$$^{238}_{92}\text{U} \rightarrow {}^{234}_{90}\text{Th} + {}^{4}_{2}\text{He}$$

Note that the mass numbers balance (238 = 234 + 4) and that the atomic numbers (the number of protons and, therefore, charge) balance (92 = 90 + 2).

The reactions and species involved are summarized below.

Decay Type	Radiation Emitted	Generic Reaction Equation
Alpha	${}^{4}_{2}\text{He}$	${}^{A}_{Z}\text{X} \rightarrow {}^{A-4}_{Z-2}\text{X}' + {}^{4}_{2}\text{He}$
Beta	${}^{0}_{-1}\text{e}$	${}^{A}_{Z}\text{X} \rightarrow {}^{A}_{Z+1}\text{X}' + {}^{0}_{-1}\text{e}$
Gamma	${}^{0}_{0}\gamma$	${}^{A}_{Z}\text{X} \rightarrow {}^{A}_{Z}\text{X}' + {}^{0}_{0}\gamma$

PROCEDURE

(1) Complete Questions 1–7.

RADIOACTIVE DECAY

The radioactive half-life for a radioisotope is the time needed for half of the radioactive nuclei in a sample to undergo radioactive decay. After one half-life, there will be one-half of the original material remaining; after two half-lives there will be one-fourth remaining, and so on. The relationship between the original number of nuclei at time zero (N_0) and the number (N) at a later time, t, is described by:

$$N = N_0 e^{-\lambda t}$$

where e is the base of the natural logarithm ($e = 2.7182$) and λ is the decay constant for the nuclide. Once we know λ, we can calculate the half-life ($t_{1/2}$).

$$\lambda = \frac{\ln(2)}{t_{1/2}} \approx \frac{0.693}{t_{1/2}}$$

PROCEDURE

(1) Complete Questions 8–13.

LAB 22 - ASSIGNMENT

QUESTIONS - RADIOACTIVE DECAY

(1) The reaction of an alpha particle with magnesium-25 $\left({}^{25}_{12}\text{Mg}\right)$ produces a proton and a daughter nuclide of another element. Show the balanced reaction and name the new nuclide produced.

(2) The nuclide ${}^{125}_{53}\text{I}$ combines with an electron and produces a new nucleus and no other particles. What is the equation for this reaction?

(3) The first naturally occurring unstable element isolated, polonium-212, decays by emitting an alpha particle. What is the equation for this reaction?

(4) The first nuclide to be prepared by artificial means was an isotope of oxygen $\left({}^{17}_{8}\text{O}\right)$. It was made by Ernest Rutherford in 1919 by bombarding nitrogen-14 with alpha particles. What is the equation for this reaction?

(5) James Chadwick discovered the neutron in 1932, as a previously unknown neutral particle produced along with carbon-12 by the nuclear reaction between berillium-9 and helium-4. What is the equation for this reaction?

(6) The first element to be prepared that does not occur naturally on the earth, technetium-97, was created by bombardment of molybdenum-97 by deuterons $\left({}^{2}_{1}\text{H}\right)$. What is the equation for this reaction?

(7) The first controlled nuclear chain reaction was carried out in a reactor at the University of Chicago in 1942. One of the reactions involved was the decay of uranium-235 into bromine-87 and Lanthanam-146. The uranium was bombarded with neutrons. What is the equation for this reaction?

QUESTIONS – RADIOACTIVE DECAY

(8) The half-life of Zn-71 is 2.4 min. If you start with 100 g at time zero, how many grams would be left after 7.2 min?

(9) Pd-100 has a half-life of 3.6 days. If we had $6.02 \cdot 10^{23}$ atoms at the start, how many atoms would be present after 20 days.

(10) Os-182 has a half-life of 21.5 hrs. How many grams of a 20 gram sample would have decayed after exactly three half-lives?

(11) After 24 days, 2 mg of an original 128 mg sample remain. What is the half-life of the sample?

(12) How long will it take for a 40.0 g of I-131 (half-life = 8.040 days) to decay to 1/100 its original mass?

(13) The half-life of Palladium-100 is 4 days. After 12 days a sample of Pd-100 has been reduced to a mass of 4 mg. What was the starting mass?

APPENDIX A - REFERENCE TABLES

Metric Prefixes

Prefix	Symbol	Multiple	Prefix	Symbol	Multiple
peta	P	10^{15}	milli-	m	10^{-3}
tera	T	10^{12}	micro-	μ	10^{-6}
giga-	G	10^{9}	nano-	n	10^{-9}
mega-	M	10^{6}	pico	p	10^{-12}
kilo-	K	10^{3}	femto-	f	10^{-15}

SI Base Units

Quantity	Unit	Quantity	Unit
Length	meter (m)	Electrical Current	ampere (A)
Mass	kilogram (kg)	Amount of Substance	mole (mol)
Time	second (s)	Luminous Intensity	candela (cd)
Temperature	kelvin (K)		

SI Defined Constants

Quantity	Symbol	Value & Units	Defining Unit
Cs-133 hyperfine frequency	$\Delta\nu_{Cs}$	9,192,631,770 Hz	defines the second
speed of light	c	$3\cdot10^{8}$ m/s	defines the meter
Planck's constant	h	$6.63\cdot10^{-34}$ J·s	defines the kilogram
elementary charge	e	$1.6\cdot10^{-19}$ C	defines the ampere
Boltzmann constant	k or k_B	$1.38\cdot10^{-23}$ J·K^{-1}	defines the kelvin
Avogadro constant	N_A	$6.02\cdot10^{23}$ mol^{-1}	defines the mole
luminous efficacy	K_{cd}	683 lm·W^{-1}	defines the candela

SI Coherent Derived Units

Quantity	Name	Symbol	Equivalent(s)	SI Base Units
frequency	hertz	Hz	1/s	s^{-1}
angle	radian	Rad	m/m	1
solid angle	steradian	sr	m^2/m^2	1
force	newton	N	$kg\cdot m/s^2$	$kg\cdot ms^{-2}$
pressure, stress	pascal	Pa	N/m^2	$kg\cdot m^{-1}\cdot s^{-2}$
energy, work, heat	joule	J	N·m, C·V, W·s	$kg\cdot m^2\cdot s^{-2}$
power, radiant flux	watt	W	J/s, V·A	$kg\cdot m^2\cdot s^{-3}$
electrical charge	coulomb	C	s·A, F·V	s·A
electric potential (voltage)	volt	V	J/C, W/A	$kg\cdot m^2\cdot s^{-3}\cdot A^{-1}$
electrical resistance	ohm	Ω	V/A, 1/S, s/F, V^2/W	$kg\cdot m^2\cdot s^{-3}\cdot A^{-2}$
electrical conductance	siemens	S	A/V, 1/Ω	$s^3\cdot A^2\cdot kg^{-1}\cdot m^{-2}$
electrical capacitance	farad	F	C/V, s/Ω	$s^4\cdot A^2\cdot kg^{-1}\cdot m^{-2}$
electrical inductance	henry	H	Ω·s, V·s/A	$kg\cdot m^2\cdot s^{-2}A^{-2}$
magnetic flux density	tesla	T	N/A·m, Wb/m^2, V·s/m^2	$kg\cdot s^{-2}\cdot A^{-1}$
magnetic flux	weber	Wb	V·s, J/A, T·m^2	$kg\cdot m^2\cdot s^{-2}A^{-1}$
luminous flux	lumen	lm	cd·sr	cd
illuminance	lux	lx	lm/m^2	$cd\cdot m^{-2}$

Other SI Constants

Quantity	Symbol	Formula*	Value & Units
atomic mass constant	m_μ	---	$1.66 \cdot 10^{-27}$ kg
electron volt	eV	---	$1.6 \cdot 10^{-19}$ J
gravitational constant	G	---	$6.67 \cdot 10^{-11}$ N·m²·kg⁻²
molar gas constant	R	---	8.31 J·mol⁻¹·K⁻¹
electric permittivity of free space	ε_0	$\frac{1}{\mu_0 c^2}$	$8.85 \cdot 10^{-12}$ F·m⁻¹
magnetic permeability of free space	μ_0	$\frac{1}{\varepsilon_0 c^2}$	$1.26 \cdot 10^{-6}$ T·m·A⁻¹
Coulomb's constant	k or k_e	$\frac{1}{4\pi\varepsilon_0}$	$8.99 \cdot 10^{9}$ N·m²·C⁻²
Stefan-Boltzmann constant	σ	$\frac{2\pi^5 k_B^4}{15h^3c^2}$	$5.67 \cdot 10^{-8}$ W·m⁻²·K⁻⁴
Rydberg constant	$R = R_\infty$	$\frac{m_e q_p^4}{8\varepsilon_0^2 h^3 c}$	$1.097 \cdot 10^{7}$ m⁻¹

Coefficients of Friction

System	μ_s	μ_k	System	μ_s	μ_k
rubber on concrete (dry)	0.90	0.68	steel on steel	0.74	0.57
rubber on concrete (wet)	0.74	0.58	Teflon on steel	0.04	0.04
rubber on asphalt (dry)	0.85	0.68	copper on steel	0.53	0.36
rubber on asphalt (wet)	0.70	0.53	aluminum on steel	0.61	0.47
bones + synovial fluid	0.01	0.003	wood on wood	0.25-0.05	0.20
glass on glass	0.94	0.40	steel on ice	0.40	0.02

Coefficients of Drag

Shape	C_d	Object	C_d
sphere	0.47	skydiver (horizontal)	1.00
cone (point to wind)	0.50	skydiver (feet first)	0.70
cube (face to wind)	1.05	vehicle – semi tractor-trailer	0.70
cube (corner to wind)	0.80	vehicle – SUV	0.64
circular flat plate	1.12	vehicle – pickup truck	0.43
airfoil	0.05	vehicle – car	0.29

Density and Viscosity (Air, Water & Seawater)

Temp (°C)	Density (ρ) kg/m³			Dynamic Viscosity (η) N·s/m²		
	Air	Water	Seawater	Air	Water	Seawater
0	1.292	999.9	1028	$1.72 \cdot 10^{-5}$	$1.79 \cdot 10^{-3}$	$1.89 \cdot 10^{-3}$
5	1.268	1000.0	1028	$1.74 \cdot 10^{-5}$	$1.55 \cdot 10^{-3}$	$1.61 \cdot 10^{-3}$
10	1.246	999.7	1027	$1.76 \cdot 10^{-5}$	$1.31 \cdot 10^{-3}$	$1.41 \cdot 10^{-3}$
15	1.225	999.1	1026	$1.79 \cdot 10^{-5}$	$1.17 \cdot 10^{-3}$	$1.22 \cdot 10^{-3}$
20	1.204	998.2	1025	$1.81 \cdot 10^{-5}$	$1.00 \cdot 10^{-3}$	$1.09 \cdot 10^{-3}$
25	1.184	997.0	1023	$1.84 \cdot 10^{-5}$	$8.90 \cdot 10^{-4}$	$9.70 \cdot 10^{-4}$
30	1.164	995.7	1022	$1.86 \cdot 10^{-5}$	$7.97 \cdot 10^{-4}$	$8.7 \cdot 10^{-4}$
50	1.093	988.0	---	$1.95 \cdot 10^{-5}$	$5.47 \cdot 10^{-4}$	---
100	0.947	958.4	---	$2.17 \cdot 10^{-5}$	$2.82 \cdot 10^{-4}$	---

Elastic Moduli

Material	Values (Pa)		
	Young's Modulus (*Y*)	Shear Modulus (*S*)	Bulk Modulus (*B*)
Aluminum	$7.0 \cdot 10^{10}$	$2.5 \cdot 10^{10}$	$7.5 \cdot 10^{10}$
Bone (tension)	$1.6 \cdot 10^{10}$	$8.0 \cdot 10^{10}$	$0.8 \cdot 10^{10}$
Bone (compression)	$0.9 \cdot 10^{10}$	$8.0 \cdot 10^{10}$	$0.8 \cdot 10^{10}$
Brass	$9.0 \cdot 10^{10}$	$3.5 \cdot 10^{10}$	$6.0 \cdot 10^{10}$
Copper	$11.0 \cdot 10^{10}$	$4.4 \cdot 10^{10}$	$14.0 \cdot 10^{10}$
Glass	$6.0 \cdot 10^{10}$	$2.5 \cdot 10^{10}$	$5.0 \cdot 10^{10}$
Granite	$4.5 \cdot 10^{10}$	$2.0 \cdot 10^{10}$	$4.5 \cdot 10^{10}$
Iron	$21.0 \cdot 10^{10}$	$7.7 \cdot 10^{10}$	$16.0 \cdot 10^{10}$
Lead	$1.6 \cdot 10^{10}$	$0.6 \cdot 10^{10}$	$4.1 \cdot 10^{10}$
Steel	$20.0 \cdot 10^{10}$	$7.5 \cdot 10^{10}$	$16.0 \cdot 10^{10}$
Water	---	---	$0.22 \cdot 10^{10}$

Coefficients of Linear (α) & Volume (β) Expansion

Solids			Liquids	
Material	α (/°C)	β (/°C)	Material	β (/°C)
Aluminum	$25 \cdot 10^{-6}$	$75 \cdot 10^{-6}$	Water	$210 \cdot 10^{-6}$
Brass	$19 \cdot 10^{-6}$	$56 \cdot 10^{-6}$	Ether	$1650 \cdot 10^{-6}$
Copper	$17 \cdot 10^{-6}$	$51 \cdot 10^{-6}$	Ethyl Alcohol	$1100 \cdot 10^{-6}$
Gold	$14 \cdot 10^{-6}$	$42 \cdot 10^{-6}$	Gasoline	$950 \cdot 10^{-6}$
Iron (steel)	$12 \cdot 10^{-6}$	$35 \cdot 10^{-6}$	Glycerin	$500 \cdot 10^{-6}$
Glass (ordinary)	$9 \cdot 10^{-6}$	$27 \cdot 10^{-6}$	Mercury	$180 \cdot 10^{-6}$
Glass (Pyrex)	$3 \cdot 10^{-6}$	$9 \cdot 10^{-6}$	Gases	$3400 \cdot 10^{-6}$
Concrete	$12 \cdot 10^{-6}$	$36 \cdot 10^{-6}$		

Moments of Inertia

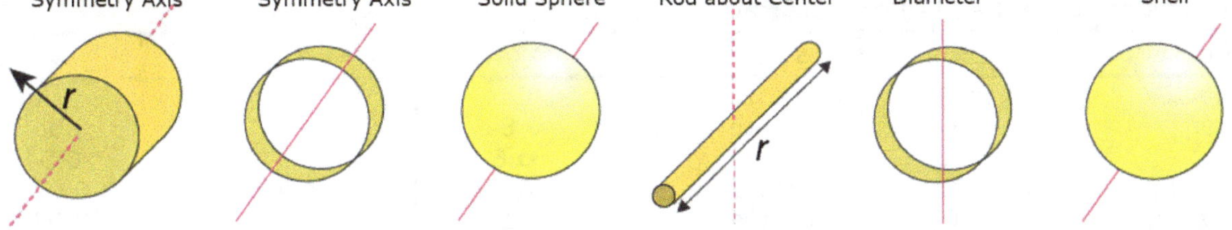

Solid Cylinder about Symmetry Axis	Hoop about Symmetry Axis	Solid Sphere	Rod about Center	Hoop about Diameter	Thin Spherical Shell
$I = \frac{1}{2}MR^2$	$I = MR^2$	$I = \frac{2}{5}MR^2$	$I = \frac{1}{12}ML^2$	$I = \frac{1}{2}MR^2$	$I = \frac{2}{3}MR^2$

Specific Heats (*c*) of Various Substances

Solids*		Liquids*		Gases**	
Substance	*c* (J/kg·°C)	Substance	*c* (J/kg·°C)	Substance	*c* (J/kg·°C)
Aluminum	900	Benzene	1740	Air (dry)	721
Copper	387	Ethanol	2450	Ammonia	1670
Glass	840	Glycerin	2410	Carbon Dioxide	638
Human Body	3500	Mercury	139	Nitrogen	739
Water - Ice (c_I)	2090	Water (c_W)	4186	Oxygen	651
Iron (steel)	452			Water - Vapor (c_V)	1860
Wood	1700				

*At constant volume and 25°C

**At constant volume and 20°C (except water at 100°C)

Heats of Fusion (L_f) & Vaporization (L_v)				
Substance	**Melting Point (°C)**	**L_f (kJ/kg)**	**Boiling Point (°C)**	**L_v (kJ/kg)**
Hydrogen	-259.3	58.6	-252.9	452
Nitrogen	-210.0	25.5	-195.8	201
Oxygen	-218.8	13.8	-183.0	213
Water	0.0	334	100	2256
Aluminum	660	380	2450	11400
Gold	1063	64.5	2660	377
Copper	1083	134	2595	1211
Tungsten	3410	184	5900	1150

Conductivity (σ), Resistivity (ρ), and Temp. Coefficients (α)			
Conductors	**σ (1/Ω·m) ***	**ρ (Ω·m) ***	**α (1/°C)**
Silver	$6.29 \cdot 10^{7}$	$1.59 \cdot 10^{-8}$	0.0038
Copper	$5.95 \cdot 10^{7}$	$1.68 \cdot 10^{-8}$	0.0039
Gold	$4.10 \cdot 10^{7}$	$2.44 \cdot 10^{-8}$	0.0034
Aluminum	$3.77 \cdot 10^{7}$	$2.65 \cdot 10^{-8}$	0.0039
Iron	$1.03 \cdot 10^{7}$	$9.71 \cdot 10^{-8}$	0.0065
Steel	$0.50 \cdot 10^{7}$	$20 \cdot 10^{-8}$	---
Mercury	$0.10 \cdot 10^{7}$	$98 \cdot 10^{-8}$	0.0009
Semi-Conductors	**σ (1/Ω·m) ***	**ρ (Ω·m) ***	**α (1/°C)**
Carbon (pure)	$2.86 \cdot 10^{-6}$	$3.50 \cdot 10^{-5}$	-0.0005
Germanium (pure)	---	0.60	-0.048
Silicon (pure)	---	2300	-0.075
Insulators	**σ (1/Ω·m) ***	**ρ (Ω·m) ***	**α (1/°C)**
Amber	$2.00 \cdot 10^{-15}$	$5 \cdot 10^{14}$	---
Quartz (fused)	$2.00 \cdot 10^{-15}$	$75 \cdot 10^{16}$	---

*σ and ρ at 20°C

Thermal Conductivities (k) of Materials (W/m·°C)					
Substance	**k**	**Substance**	**k**	**Substance**	**k**
Diamond	2000	Glass (average)	0.84	Snow (dry)	0.1
Copper	390	Concrete/Brick	0.84	Cork	0.042
Gold	318	Water	0.6	Wool	0.04
Aluminum	220	Fatty Tissue	0.2	Down feathers	0.025
Iron (steel)	80	Asbestos	0.16	Air	0.023
Steel (stainless)	14	Plasterboard	0.16	Polystyrene foam	0.01
Ice	2.2	Wood (average)	0.12		

Dielectric Constant (κ) and Strength (E_c)					
Material	**κ**	**E_c (10^6 V/m)**	**Material**	**κ**	**E_c (10^6 V/m)**
Vacuum	1.0	∞	Fused Quartz	3.78	8
Dry Air (1 atm)	1.0006	3.0	Mica	5.4	100 – 200
Teflon	2.1	60 – 173	Pyrex Glass	5.6	2000
Polyethylene	2.25	19 – 160	Neoprene Rubber	6.7	118
Paraffin	2.3	11	Beryllium Oxide	6.7	27 – 31
Polystyrene	2.56	19.7	Calcium Oxide	11.8	> 350
Silicone Rubber	3.2	10 – 30	Acetone	20.7	< 10
Nylon	3.4	14	Water	80	15.7-26.7
Paper	3.7	16			

Properties of Water

Physical Properties		Vapor Pressure vs Temperature	
Property	**Value**	**Temp (°C)**	**Vapor Pressure (Pa)**
Density (4°C)	1000 kg/m^3	0	610.5
Density (0°C) Ice	917 kg/m^3	5	872.3
Density (100°C) Steam	0.6 kg/m^3	10	1228
Density (Sea Water)	1025 kg/m^3	15	1705
Latent Heat of Fusion (L_f)	334 kJ/kg	20	2338
Latent Heat of Vaporization (L_v)	2256 kJ/kg	23	2809
Specific Heat (c_i) – ice	2.09 kJ/kg	25	3167
Specific Heat (c_w) – water	4.186 kJ/kg	30	4243
Specific Heat (c_v) - vapor	1.86 kJ/kg	35	5623
Thermal Conductivity	0.6 J/s·m·°C	40	7376

Inner Solar System Data

	Earth-Sun-Moon System			Inner Planets		
	Earth	**Sun**	**Moon**	**Mercury**	**Venus**	**Mars**
Mass (10^{24} kg except Sun)	5.97	$1.99 \cdot 10^{30}$	0.073	0.33	4.87	0.642
Radius (10^6 m)	6.37	695.7	1.74	2.44	6.05	3.40
Gravity (m/s^2)	9.8	274	1.62	3.7	8.87	3.71
Density (kg/m^3)	5514	1408	3340	5427	5243	3933
Distance from Sun (10^6 km)	149.6	---	0.384*	57.9	108.2	227.9
Perihelion (10^6 km)	147.1	---	0.363*	46.0	107.5	206.6
Aphelion (10^6 km)	152.1	---	0.406*	69.8	108.9	249.2
Orbital Period (days)	365.2	---	27.2*	88.0	224.7	687.0
Rotational Period (hours)	23.9	---	655.7	1407.6	-5832.5	24.6

*Relative to Earth (not Sun)

Outer Solar System Data

	Outer Planets				Dwarf Planets	
	Jupiter	**Saturn**	**Uranus**	**Neptune**	**Pluto**	**Ceres**
Mass (10^{24} kg)	1898	568	86.8	102	0.015	0.00094
Radius (10^6 km)	71.49	60.27	25.56	24.76	1.20	0.476
Gravity (m/s^2)	23.1	9.0	8.7	11.0	0.7	0.27
Density (kg/m^3)	1326	687	1271	1638	2095	2150
Distance from Sun (10^6 km)	778.6	1433.5	28.72.5	4495.1	5906.4	413.8
Perihelion (10^6 km)	740.5	1352.6	2741.3	4444.5	4436.8	381.9
Aphelion (10^6 km)	816.6	1514.5	3003.6	4545.7	7375.9	446.1
Orbital Period (days)	4331	10,747	30,589	59,800	90,560	1680
Rotational Period (hrs)	9.9	10.7	-17.2	16.1	-153.3	9.07

Greek Alphabet

Name	Upper	Lower	Name	Upper	Lower	Name	Upper	Lower
Alpha	A	A	Iota	I	ι	Rho	P	ρ
Beta	B	β	Kappa	K	κ	Sigma	Σ	σ
Gamma	Γ	γ	Lambda	Λ	λ	Tau	T	τ
Delta	Δ	δ	Mu	M	μ	Upsilon	Y	υ
Epsilon	E	ε	Nu	N	ν	Phi	Φ	ϕ
Zeta	Z	ζ	Xi	Ξ	ξ	Chi	X	χ
Eta	H	η	Omicron	O	ο	Psi	Ψ	ψ
Theta	Θ	θ	Pi	Π	π	Omega	Ω	ω

Trigonometry (Right Angle)

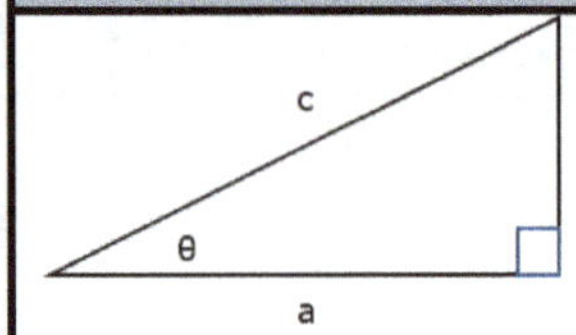

$$\sin\theta = \frac{b}{c}$$
$$\cos\theta = \frac{a}{c}$$
$$\tan\theta = \frac{\sin\theta}{\cos\theta} = \frac{b}{a}$$
$$\cot\theta = \frac{\cos\theta}{\sin\theta} = \frac{1}{\tan\theta} = \frac{a}{b}$$
$$\sec\theta = \frac{1}{\cos\theta} = \frac{c}{a}$$
$$\csc\theta = \frac{1}{\sin\theta} = \frac{c}{b}$$

$$\sin^2(\theta)+\cos^2(\theta)=1$$
$$1+\tan^2(\theta)=\sec^2(\theta)$$
$$1+\cot^2(\theta)=\csc^2(\theta)$$
$$\sin(2\theta)=2\sin(\theta)\cos(\theta)$$
$$\cos(2\theta)=\cos^2(\theta)-\sin^2(\theta)=1-2\sin^2(\theta)$$
$$\tan(2\theta)=\frac{2\tan(\theta)}{1-\tan^2(\theta)}$$

Subatomic Particles

Particle	Symbol	Mass (kg)	Charge (C)	Relative Charge
Electron	e^-	$m_e = 9.11 \cdot 10^{-31}$	$q_e = -1.6 \cdot 10^{-19}$	-1
Proton	p^+	$m_p = 1.67 \cdot 10^{-27}$	$q_p = +1.6 \cdot 10^{-19}$	+1
Neutron	n	$m_n = 1.67 \cdot 10^{-27}$	---	0

Standard Model of Elementary Particles

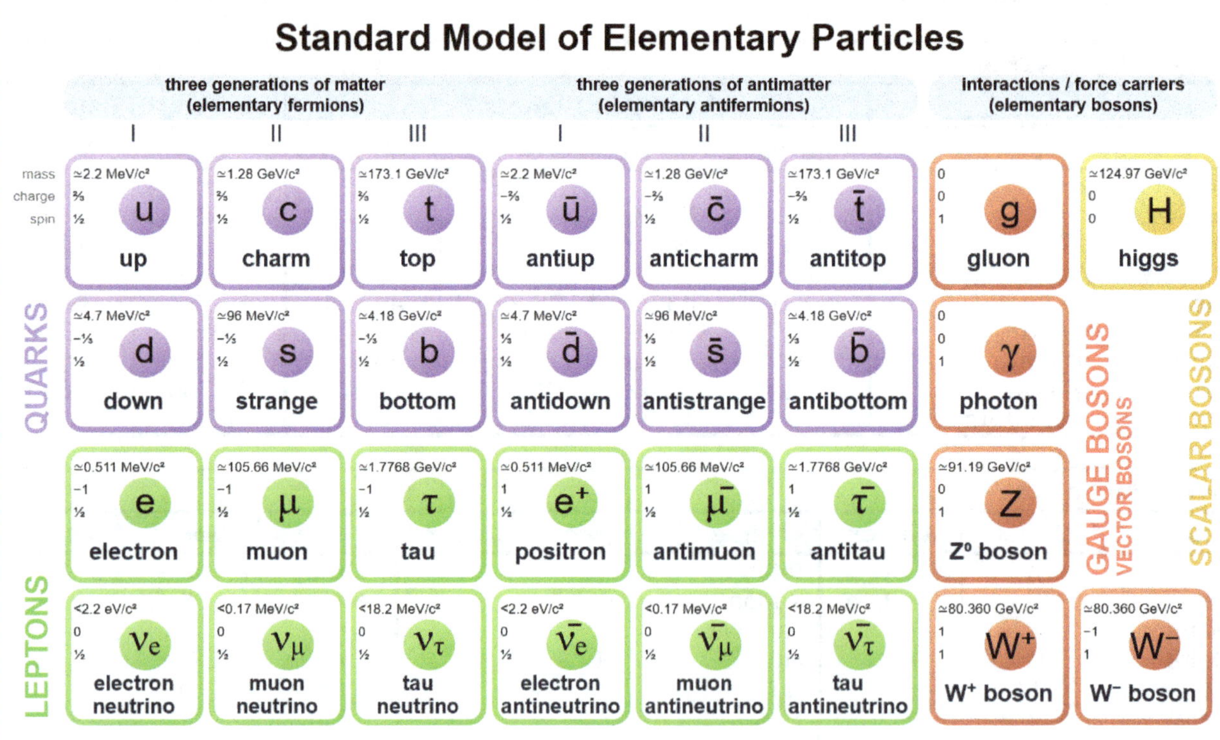

www.ingramcontent.com/pod-product-compliance
Lightning Source LLC
LaVergne TN
LVHW082337200926
848481LV00012B/3374
9798319729033